LAND-USE CHANGES AND THEIR ENVIRONMENTAL IMPACT IN RURAL AREAS IN EUROPE

MAN AND THE BIOSPHERE SERIES

MAN AND THE BIOSPHERE SERIES

Series Editor J.N.R. Jeffers

VOLUME 24

LAND-USE CHANGES AND THEIR ENVIRONMENTAL IMPACT IN RURAL AREAS IN EUROPE

Edited by

R. Krönert
UFZ, Leipzig, Germany

J. Baudry
Institut National de la Recherche Agronomique, Rennes, France

I. R. Bowler
University of Leicester, UK

A. Reenberg
University of Copenhagen, Denmark

PUBLISHED BY

PARIS

AND

The Parthenon Publishing Group
International Publishers in Science, Technology & Education

Published in 1999 by the United Nations Educational, Scientific and Cultural Organization,
7 Place de Fontenoy, 75700 Paris, France—UNESCO ISBN 92-3-103596-7
and
The Parthenon Publishing Group Inc.
One Blue Hill Plaza
PO Box 1564, Pearl River,
New York 10965, USA—ISBN 1-85070-047-8
and
The Parthenon Publishing Group Limited
Casterton Hall, Carnforth,
Lancs LA6 2LA, UK—ISBN 1-85070-047-8

© Copyright **UNESCO 1999**

British Library Cataloguing in Publication Data
Land-use changes and their environmental impact in rural areas in Europe – (Man and the
biosphere ; v. 24)
1. Land use, Rural – Europe 2. Land use, Rural – Environmental aspects – Europe
I. Krönert, R.
333.7'6'094
ISBN 1-85070-047-8

Library of Congress Cataloging-in-Publication Data
Land-use changes and their environmental impact in rural areas in Europe / edited by R.
Krönert ... [et al.], (editors).
    p.   cm. — (Man and the biosphere series ; v. 24)
Includes bibliographical references.
ISBN 1-85070-047-8
    1. Land use. Rural—Europe. 2. Land use. Rural—Environmental aspects—Europe.
    I. Krönert, Reinhard. II. Series.
HD586.L3358   1999
333.76'13'094—dc21   98-13170
                                                                         CIP

Typeset by Speedlith Photolitho Ltd., Manchester, UK
Printed and bound by Butler & Tanner Ltd., Frome and London, UK

CONTENTS

PREFACE   vii

LIST OF CONTRIBUTORS   xi

1.   INTRODUCTION   1
      R. Krönert

2.   LAND-USE AND LANDSCAPE CHANGES – THE   23
      CHALLENGE OF COMPARATIVE ANALYSIS OF
      RURAL AREAS IN EUROPE
      A. Reenberg and J. Baudry

3.   DESCRIPTION AND ANALYSIS OF THE NATURAL   43
      RESOURCE BASE
      O. Bastian

4.   LAND USE, NATURE CONSERVATION AND   65
      REGIONAL POLICY IN ALENTEJO, PORTUGAL
      F. Bacharel and T. Pinto-Correia

5.   RURAL LAND-USE AND LANDSCAPE DYNAMICS –   81
      ANALYSIS OF 'DRIVING FORCES' IN SPACE
      AND TIME
      J. Brandt, J. Primdahl and A. Reenberg

6.   DRIVING FACTORS OF LAND-USE DIVERSITY AND   103
      LANDSCAPE PATTERNS AT MULTIPLE SCALES –
      A CASE STUDY IN NORMANDY, FRANCE
      J. Baudry, C. Laurent, C. Thenail, D. Denis and F. Burel

7. AGRICULTURAL LAND-USE AND LANDSCAPE CHANGE UNDER THE POST-PRODUCTIVIST TRANSITION – EXAMPLES FROM THE UNITED KINGDOM 121
I.R. Bowler and B.W. Ilbery

8. CHANGES OF RURAL LAND USE WITHIN AN AGGLOMERATION – LEIPZIG–HALLE AS AN EXAMPLE 141
R. Krönert

9. LANDSCAPE CHANGES IN ESTONIA – CAUSES, PROCESSES AND CONSEQUENCES 165
Ü. Mander and H. Palang

10. LAND-USE CHANGE IN THE AGRICULTURAL REGION OF WIELKOPOLSKA, POLAND 189
L. Ryszkowski and S. Bałazy

11. LAND-USE/LAND-COVER CHANGES UNDER AGRICULTURAL IMPACTS AND THE PROSPECTS OF ECOLOGICAL AGRICULTURAL DEVELOPMENT IN THE EUROPEAN PART OF RUSSIA 205
E.V. Milanova, V.N. Solntsev, P.A. Tcherkashin, E. Yu. Lioubimtseva and N.N. Kalutskova

12. ECOLOGICAL–ECONOMIC PROBLEMS OF LAND USE IN THE UKRAINE 221
V. Voloshyn, S. Lisovsky, I. Gukalova and V. Reshetnik

13. QUALITATIVE CHANGES IN NATURAL LANDSCAPES AND LAND USE FOLLOWING RADIOACTIVE POLLUTION FROM THE CHERNOBYL ACCIDENT 235
V. Davydchuk

14. CONCLUSIONS 245
R. Krönert

INDEX 249

PREFACE

UNESCO's Man and the Biosphere Programme

Improving scientific understanding of the natrual and social processes relating to human interactions with the environment; providing information useful to decision-making on resource use; promoting the conservation of genetic diversity as an integral part of land management; enjoining the efforts of scientists, policy-makers, and local people in problem-solving ventures; mobilizing resources for field activities; strengthening of regional cooperative frameworks – these are some of the generic characteristics of UNESCO's Man and the Biosphere (MAB) Programme.

The MAB Programme was launched in the early 1970s. It is a nationally based, international programme of research, training, demonstration, and information diffusion. The overall aim is to contribute to efforts for providing the scientific basis and trained personnel needed to deal with problems of rational utilization and conservation of resources and resource systems and problems of human settlements. MAB emphasizes research for solving problems. It thus involves research by interdisciplinary teams on the interactions among ecological and social systems, field training, and application of a systems approach to understanding the relationships among the natural and human components of development and environmental management.

MAB is a decentralized programme with field projects and training activities in all regions of the world. These are carried out by scientists and technicians from universities, academies of sciences, national research laboratories, and other research and development institutions under the auspices of more than one hundred MAB National Committees. Activities are undertaken in cooperation with a range of international governmental and non-governmental organizations.

Man and the Biosphere Book Series

The Man and the Biosphere Book Series was launched to communicate some of the results generated by the MAB Programme and is aimed primarily at upper level university students, scientists, and resource managers, who are not necessarily specialists in ecology. The books are not normally suitable for undergraduate text books but rather provide additional resource material in the form of case studies based on primary data collection and written by the researchers involved; global and regional syntheses of comparative research conducted in several sites or countries; and state-of-the-art assessments of knowledge or methodological approaches used on scientific meetings, commissioned reports, or panels of experts.

Land-use changes in Europe

The 1989–1990 period saw a complete upheaval in the post Second World War international order, particularly in Europe. It was a period of unprecedented political, economic and social change and, in Europe, these changes were reflected in the changing shape of scientific co-operation in the region. In turn, the moving landscape of scientific relationships was reflected in microcosm within the MAB Programme, and more particularly in proposals for new co-operative studies among European countries.

Impetus for a new generation of collaborative work was provided by the Second All-European (including North America) meeting of MAB National Committees which took place in May 1989 in Třebon (Czechoslovakia), following an earlier conference held at Berchtesgaden (Federal Republic of Germany) in 1987.

The Třebon conference focussed on the mid-latitude temperate regions. Separate working groups were set up to examine future co-operatioin in four particular areas: biosphere reserves, ecotones, forest ecosystems and land-use change. Discussions also took place on three other topics: mountain ecosystems, urban systems and environmental education.

Each of these working groups based their discussions on proposals submitted by MAB National Committees, which provided suggestions on issues that could be tackled in collaborative ventures as well as indications concerning specific field projects which might contribute to collaborative programmes in particular fields. Recommendations and proposals drawn up by working groups, and subsequently adopted by the conference, included: the launching of comparative studies on landscape pattern dynamics in European rural areas; new directions for work on temperate forest ecosystems and their response to stress and disturbance; further planning of the comparative study on land–inland water ecotones and their role in landscape management and restoration; measures for further implementation of the Action Plan for Biosphere Reserves in the European region.

Subsequently, four seminars were held within the framework of the comparative study on land-use changes in Europe, in Kiev (Ukraine) and in Caen (France) 1991, in Poznań (Poland) in October 1992 and in Bad Lauchstadt (Germany) in September 1993. Among other aims, these seminars provided an opportunity for the presentation and discussion of the findings from field research undertaken within the context of national research projects. They also provided for the examination of approaches to cross-country comparison, including the analysing of landscape pattern dynamics at a range of scales and the effects of land-use change on such pattern dynamics.

In publishing this volume in the Man and the Biosphere Series, UNESCO would like to thank volume editors Rudolph Krönert, Jacques Baudry, Ian Bowler and Anette Reenberg for planning the volume and seeing it through to publication. The cover montage is by Ivette Fabbri, based on photographs by Yann Arthus-Bertrand/Earth From Above/UNESCO and by Ivette Fabbri.

MAN AND THE BIOSPHERE SERIES

1. The Control of Eutrophication of Lakes and Reservoirs. S.-O. Ryding; W. Rast (eds.), 1989.
2. An Amazonian Rain Forest. The Structure and Function of a Nutrient Stressed Ecosystem and the Impact of Slash-and-Burn Agriculture. C.F. Jordan, 1989.
3. Exploiting the Tropical Rain Forest: An Account of Pulpwood Logging in Papua New Guinea. D. Lamb, 1990.
4. The Ecology and Management of Aquatic–Terrestrial Ecotones. R.J. Naiman; H. Décamps (eds.), 1990.
5. Sustainable Development and Environmental Management of Small Islands. W. Beller; P. d'Ayala; P. Hein (eds.), 1990.
6. Rain Forest Regeneration and Management: A. Goméz-Pompa; T.C. Whitmore; M. Hadley (eds.), 1991.
7. Reproductive Ecology of Tropical Forest Plants. K. Bawa; M. Hadley (eds.), 1990.
8. Biohistory: The Interplay between Human Society and the Biosphere – Past and Present. S. Boyden, 1992.
9. Sustainable Investment and Resource Use: Equity, Environmental Integrity and Economic Efficiency. M.D. Young, 1992.
10. Shifting Agriculture and Sustainable Development: An Interdisciplinary Study from North-Eastern India. P.S. Ramakrishnan, 1992.
11. Decision Support Systems for the Management of Grazing Lands: Emerging Issues. J.W. Stuth; B.G. Lyons (eds.), 1993.
12. The World's Savannas: Economic Driving Forces, Ecological Constraints and Policy Options for Sustainable Land Use. M.D. Young; O.T. Solbrig (eds.), 1993.
13. Tropical Forests, People and Food: Biocultural Interactions and Applications to Development. C.M. Hladik; A. Hladik; O.F. Linares; H. Pagezy; A. Semple; M. Hadley (eds.), 1993.
14. Mountain Research in Europe: An Overview of MAB Research from the Pyrenees to Siberia. M.F. Price, 1995.
15. Brazilian Perspectives on Sustainable Development of the Amazon Region. M. Clüsener-Godt; I. Sachs (eds.), 1995.
16. The Ecology of the Chernobyl Catastrophe: Scientific Outlines of an International Programme of Collaborative Research. V.K. Savchenko, 1995.
17. Ecology of Tropical Forest Tree Seedlings. M.D. Swaine (ed.), 1996.
18. Biodiversity in Land–Inland Water Ecotones. J. Lachavanne; R. Juge (eds.), 1997.
19. Population and Environment in Arid Regions. J. Clarke; D. Noin (eds.), 1998.
20. Forest Biodiversity Research, Monitoring and Modeling. F. Dallmeier; J.A. Comiskey (eds.), 1998.
21. Forest Biodiversity in North, Central and South America, and the Caribbean. F. Dallmeier; J.A. Comiskey (eds.), 1998.
22. Commons in a Cold Climate – Coastal Fisheries and Reindeer Pastoralism in North Norway: The Co-management Approach. S. Jentoft (ed.), 1998.
23. Assessment and Control of Nonpoint Source Pollution of Aquatic Ecosystems. A Practical Approach. J. Thornton; W. Rast; M. Holland; G. Jolanki; S.-O. Ryding (eds.), 1999.

LIST OF CONTRIBUTORS

Fatima Bacharel
Regional Planning Commission of
  Alentejo
Estrada das Piscinas 193
P-7000 Évora, Portugal
E-mail: fbacharel@ccr-alt.pt

Stanislaw Bałazy
Research Center for Agricultural
  and Forest Environment
Polish Academy of Sciences
Bukowska Street 19
60809 Poznań, Poland
E-mail: ryszagro@man.poznan.pl

Olaf Bastian
Sächsische Akademie der
  Wissenschaften
Arbeitsgruppe Naturhaushalt und
  Gebietscharakter
Neustädter Markt 19 (Blockhaus)
D-01097 Dresden, Germany
E-mail: bastian@rcr.urz.tu-dresden.de

Jacques Baudry
Institut National de la Recherche
  Agronomique
Unité SAD-Armorique
65, rue de Saint Brieuc
35042 Rennes Cedex, France
E-mail: jbaudry@roazhon.inra.fr

Ian R. Bowler
Department of Geography
University of Leicester
Leicester
LE1 7RH, UK
E-mail: irb@le.ac.uk

Jesper Brandt
Department of Geography and
  International Development
  Studies
Roskilde University
House 19.2, P.O. Box 260
DK-4000 Roskilde, Denmark
E-mail: Brandt@geo.ruc.dk

Françoise Burel
CNRS-Université de Rennes 1
UMR ECOBIO
Campus de Beaulieu
35042 Rennes Cedex, France

Vassili Davydchuk
Institute of Geography
National Academy of Sciences of
  Ukraine
44 Volodimirska str.
252003 Kiev, Ukraine
E-mail:
  Chornob@geogr.freenet.kiev.ua

Daniel Denis
Institut National de la Recherche
  Agronomique
Unité SAD-Armorique
65 rue de Saint Brieuc
35042 Rennes Cedex, France

Irina Gukalova
Institute of Geography
National Academy of Sciences of
  Ukraine
44 Volodymyrska St
252034 Kiev, Ukraine
E-mail: ira@iatp.kiev.ua

Brian W. Ilbery
Division of Geography
Coventry University
Priory Street
Coventry CV1 5FB, UK
E-mail: b.ilbery@coventry.ac.uk

N.N. Kalutskova
Moscow State University
Faculty of Geography
119899, Moscow, Russia
E-mail: nnk@global.geogr.msk.su

Rudolf Krönert
UFZ-Centre for Environmental
  Research Ltd.
P.O. Box 2
D-04301 Leipzig, Germany
E-mail: kroenert@alok.ufz.de

Catherine Laurent
Institut National de la Recherche
  Agronomique
Unité SAD - lle-de-France
16, rue Claude Bernard
75231 Paris Cedex 05, France

E. Yu. Lioubimtseva
Moscow State University
Faculty of Geography
119899, Moscow, Russia
E-mail: lioubimt.mangala@skynet.be

Sergey Lisovsky
Institute of Geography
National Academy of Sciences of
  Ukraine
44 Volodymyrska St
252034 Kiev, Ukraine
E-mail: root@geogr.freenet.kiev.ua

Ü. Mander
Institute of Geography
University of Tartu
Vanemuise 46
EE 2400 Tartu, Estonia
E-mail: ylo@math.ut.ee

E.V. Milanova
Faculty of Geography
M.V. Lomonosov Moscow State
  University
Vorobiovy Gory
119899 Moscow, Russia
E-mail: elena@isc.msk.su

Hannes Palang
Institute of Geography
University of Tartu
Vanemuise 46
51014 Tartu, Estonia
E-mail: hannes@ut.ee

Teresa Pinto-Correia
University of Évora
P-7000 Évora, Portugal
E-mail: mtpc@uevora.pt

Jørgen Primdahl
Dept. of Economy and Natural
  Resources
Danish Agricultural University
Rolighedsvej 23
DK-1958 Frederiksberg C, Denmark
E-mail: jpr@kvl.dk

Anette Reenberg
Institute of Geography
University of Copenhagen
Oster Voldgade 10
DK-1350 Copenhagen, Denmark
E-mail: ar@geog.ku.dk

Viktoriya Reshetnik
Institute of Geography
National Academy of Sciences of
  Ukraine
44 Volodymyrska St
252034 Kiev, Ukraine
E-mail: resh@isgeo.kiev.ua

Lech Ryszkowski
Research Center for Agricultural
  and Forest Environment
Polish Academy of Sciences
Bukowska Street 19
60809 Poznań, Poland
E-mail: ryszagro@man.poznan.pl

V.N. Solntsev
Moscow State University
Faculty of Geography
119899, Moscow, Russia
E-mail: soln@global.geogr.msu.su

P.A. Tcherkashin
Moscow State University
Faculty of Geography
119899, Moscow, Russia
E-mail: pavel@actis.ru

Claudine Thenail
Institut National de la Recherche
  Agronomique
Unité SAD - Armorique
65 rue de Saint Brieuc
35042 Rennes Cedex, France

Valentyn Voloshyn
Presidium of National Academy of
  Sciences of Ukraine
54 Volodymyrska St
252601 Kiev, Ukraine
E-mail: spop@nas.gov.ua

CHAPTER 1

INTRODUCTION

R. Krönert

THE MAB PROJECT – BACKGROUND AND OBJECTIVES

The research project 'Land-use changes in Europe and their impact on the environment – comparisons of landscape dynamics in European rural areas' was set up in May 1989, at the EUROMAB conference in Třebon, Czechoslovakia, as part of the MAB Programme. Professor Anette Reenberg (Denmark) took responsibility for coordinating the project and, with the help of J. F. Turenne (UNESCO), J. Baudry (France), myself and (initially) H. Lieth (Germany), developed its basic research framework.

The aim was to bring together scientists from both western and eastern Europe, to investigate, discuss and compare the current condition and development of Europe's rural areas. In doing so, both common features and differences across Europe were to be identified. Groups from Denmark, Estonia, France, Germany, Britain, Poland, Portugal, Russia, the Ukraine and (at times) the Czech Republic and Slovakia participated in the project. There were some clear fundamental differences in the socioeconomic conditions prevailing in Europe at the start of the work. Western Europe was (and continues to be) dominated by market–orientated, family farm agriculture, with some large enterprises. Agriculture in eastern and southern Europe was dominated by state-controlled, producer cooperatives (collectives) and state farms. Agriculture in Poland was an exception to this generalisation, since it had a much higher proportion of family farm enterprises than other communist countries.

Four seminars were held during the course of the project: in Kiev (Ukraine) and in Caen (France) in 1991, in Poznań (Poland) in 1992, and in Bad Lauchstädt near Halle (Germany) in 1993. At the 1993 seminar the project members decided to prepare a joint publication, thus providing a vehicle for further development of the work. The seminars were supported by UNESCO, as well as by national organisations and ministries. The actual research work, however, took place within the context of national research programmes. This meant that agreement on the basic principles underpinning research approaches depended heavily on the commitment of the participating groups, and the possibilities available to them in their own countries.

The seminar proceedings were published as the following (only a small number of copies were produced):

Comparisons of Landscape Pattern Dynamics in European Rural Areas. 1991 seminars, EUROMAB Research Programme, Lieury, Rennes, Kiev 1992 (eds: J. Baudry, F. Burel and V. Hawrylenko) 355pp.

1

Functional Appraisal of Agricultural Landscape in Europe. EUROMAB and INTECOL seminar 1992, Poznań, 1994 (eds: L. Ryszkowski and S. Bałazy) 307pp.

Analysis of Landscape Dynamics – Driving Factors Related to Different Scales. EUROMAB Research Programme Vol. 3, 1993 Oct. seminar, Leipzig, 1994 (ed.: R. Krönert) 164pp.

All four seminars dealt with fundamental aspects concerning the development of all European rural areas, but specific regional issues were also covered. At the Kiev seminar (local organiser, V. Voloshyn) attention was focused on the Ukrainian project work, particularly on development problems in the Carpathians and black-earth soil zones, but also on the general environmental problems in the country as a whole. In Caen (local organiser, J. Baudry) the development of the Pays d'Auge was explored in detail and the role of family farms in this development were clarified. The Poznań seminar (led by L. Ryszkowski) was held jointly with the INTECOL Agrosystems Working Group. At this seminar considerable time was devoted to a discussion of flow of material in the agricultural landscape. At the fourth seminar (local organiser, R. Krönert) the driving forces behind landscape dynamics, as related to different scales, dominated the discussion. All four seminars witnessed an exchange of knowledge about the relationship between man and his rural environment. The results should be taken into account in future scientific work and political decision making and this introduction includes a short summary of those insights, results and ideas that need to be further examined and clarified.

The radical changes taking place in European agriculture provided the background to project work. These changes are expressed primarily in terms of intensification, with increasing production per hectare of agricultural land. The consequences of this intensification are agricultural surpluses. At the same time, particularly in the more marginal areas, extensification is taking place; land may be taken out of production permanently or temporarily set aside. There are, however, regional differences within Europe, both in terms of the conditions driving such developments and the development processes themselves. Accordingly, one of the main aims of the project was to investigate the driving forces behind land-use changes under different natural and socioeconomic conditions. Other aims included the examination of the stability and sustainability of land-use systems, and the impacts of these systems on, for example, biodiversity, water pollution and soil erosion. Specific attention was also given to the intensification and marginalisation of agricultural land use, processes which are taking place simultaneously in Europe. The research approach was holistic, but took account of regional differences. There was general agreement that the work could cover a range of spatial dimensions, from case studies of small areas up to national overviews, but that middle-sized spatial units of between ten and several hundred square kilometres should be preferred. It was hoped that studies could be undertaken on landscape developments (and their causes) over relevant periods of time, since each rural area, and each landscape, contains elements defined by the past. It was also hoped that comparative studies of different agricultural regions within Europe could be undertaken. This only proved possible in one case (exemplary work comparing Denmark and Portugal). A list of parameters to be studied was finally put together which did full justice to the holistic approach taken and which allowed an in-depth study of the dynamic relationship between man and environment in rural areas (Reenberg, 1992).

Different authors have tried to produce schematic illustrations of the interactions, dynamics and driving forces behind rural areas. Bowler and Ilbery (1992) drew up a conceptual framework for researching the environmental impacts of agricultural industrialisation (Figure 7.1) in order to show the effects of industrial agriculture on the landscape. An alternative suggested by Reenberg and Pinto-Correia (1994) puts the agricultural enterprise in a more socioeconomic and landscape context (see Reenberg and Baudry, Chapter 2). It was not possible to reach common agreement on any one such schematic representation before the end of the project, due, at least in part, to the complicated nature of the topic. There was general acceptance of the representation drawn up by Messerli and Messerli (1978) from their MAB work (quoted in Milanova *et al.,* 1994b). This representation characterises the relationship between the natural complex, land use and the socioeconomic complex within the dynamic relationship between man and environment. The schematic representation given by Reenberg and Baudry (Chapter 2) builds on this approach, but places particular emphasis on aspects of this MAB project.

The agricultural enterprise is certainly the central institution responsible for preserving or changing the state of the agricultural landscape. Within the limits set by his personal rights of disposition and available capital, the owner, tenant or manager of the enterprise acts in response to:

(1) The demands of national and international markets;
(2) conditions determined by national and international agricultural policies;
(3) relevant technical developments; and
(4) natural production conditions.

Socioeconomic and political environments prove to be crucial in determining whether agricultural production and landscape manipulation cause the environmental carrying capacity of the site in question to be exceeded (overexploitation), leading to environmental damage. Soil compaction, soil erosion, an increasing loss of biodiversity, and pollution of ground and surface waters with nitrogen complexes and phosphates (as a result of too much fertiliser use) are common features of European agricultural regions. The implication is clear; socioeconomic conditions for agricultural production in Europe are less than optimal.

THE DRIVING FORCES BEHIND AGRICULTURAL LANDSCAPE DEVELOPMENT

There was general agreement among project members that the key driving forces behind changes in rural regions and agricultural landscapes are national and international markets for agricultural products, agricultural policies, technical developments and the subsequent responses of the farmer (whether owner, tenant or manager). The importance of the natural landscape, traditional land-use patterns and enterprise structures varies in each region. 'The main proposition is that the dominant model of 'industrialized agriculture' has emerged over the last four decades in capitalist economies, with damaging environmental consequences for rural areas. The 'industrial model' for agriculture is driven by international and national processes of an economic and political character, with localized results for the rural environment; nevertheless, the environmental impacts vary in nature and intensity from locality to locality. A number of comparative locality studies are required so as to identify the causes of the varying characters of agricultural industrialisation and its environmental impacts' (Bowler and Ilbery, 1992). This industrialisation was also well advanced in communist countries, where large enterprises were dominant and where there was often complete separation of crop and livestock enterprises. In the EU, the development of agricultural regions is currently taking place against a background of over-production and varying degrees of environmental damage caused by an agriculture which is still heavily subsidised. In the former communist countries, the last few years have seen a transition to a market economy, with the break up of collective and state farms (the *kolkhoz* and *sovkhoz* of the former Soviet Union). The consequent decrease in agricultural production has been accelerated by the loss of export opportunities which followed the reduction in agricultural subsidies. Although the environmental damage caused by agriculture in these countries has also decreased, it is still high.

Agricultural development trends differ, therefore, between western, eastern (and southern) Europe. They also vary considerably between countries within a regional group, and between regions within each country. The MAB project did not contain representatives from all European countries, let alone all local regions. Nevertheless, one of the aims of the cooperation was to make it clear, first, that regional aspects must be accounted for more directly when considering the development of agricultural areas; and, second, that a pan-European agricultural policy approach will have different impacts in different regions, not all of which will be desirable.

The basic principles of agricultural policy, as pursued by the EU since the reforms of 1992, are presented by Bowler and Ilbery in Chapter 7. Bauer (1994) has been very critical of these agricultural reforms. He suggests that the reforms are just a short-term compromise that does not solve the problems of over-production and environmental damage. He also concludes that:

(1) The reforms lead to additional bureaucracy and administration within the agricultural policy system;

(2) the huge direct transfer payments are just used as a means of correcting existing income distribution inequities and there is no orientation toward generally accepted social or ecological criteria;

(3) the set aside component of the reforms makes little sense from an economic or ecological point of view;

(4) the reforms, particularly the compensation payments, will lead to some stabilisation in marginal areas;

(5) the expected reduction or removal of these payments will, however, cause significant land-use change in marginal areas. Agriculture will then withdraw to those regions with good natural conditions.

According to Bauer then, the reforms do not offer a long-term solution to the problems of marginal areas.

DEVELOPMENT PHASES AND CHANGES IN AGRICULTURAL REGIONS

The Dutch study 'Ground for Choices', as described by Ivan Latestejn (1994), has been met with a great deal of interest, though by no means with total acceptance. This study assessed the long-term changes in land use associated with each of four (theoretical) development scenarios for EU agriculture and forestry:

(1) An international free market for agricultural products, with minimum interest in social considerations and no concern for the environment;

(2) development policy aimed at minimising changes in the current distribution of labour;

(3) nature and landscape conservation involving protection of natural habitats through the exclusion of arable production from certain areas. This would lead to a decrease in the area of land available for agricultural production;

(4) integrated environmental protection through the withdrawal of particular agricultural practices, such as intensive pesticide use.

The study is not intended as a means of predicting the real future, but is rather an aid to the discussion of regional policy guidelines.

All scenarios lead to a reduction in agricultural employment. The best policy measures should, of course, successfully promote environmentally-friendly farming. The way in which agriculture is financed would have to be adjusted so that it fits European nature conservation principles. The EU has studied various land-use scenarios in 12 member states; in all cases there are considerable surpluses of agricultural land. The amount of surplus, and its regional distribution, differ from one scenario to another, but the general conclusion is always the same. This means that any policy designed to maintain the long-term use of land for agricultural purposes will meet increasing resistance. The costs of such a policy may rise sharply and the eventual results will sometimes be incompatible with other goals (e.g. nature conservation and other environmental objectives). It remains unclear whether, and for how long, subsidies can prevent the decrease of agricultural land. It is also unclear whether industrial crops are able to offer an economically efficient land-use alternative. The preservation of the mainly family farm enterprise structure and the protection of existing agricultural land from forest encroachment are, however, still stated aims of current EU agricultural policy.

This volume contains several contributions which describe the historical development of rural areas and agricultural landscapes since the Middle Ages (Ryszkowski and Bałazy, Chapter 10) or during this century (Brandt *et al.*, Chapter 5; Mander and Palang, Chapter 9). Historical developments featured significantly in the project seminars. It is possible to draw the general conclusion that movement between phases of slow development with few changes, phases of accelerated development, and phases featuring massive change and sudden developments can take place within a few years. In the Middle Ages agricultural crises, epidemics and wars represented development phases that led to the disappearance of villages and fields in central Europe. The nineteenth century saw a rapid increase in agricultural production in many parts of Europe and a revitalisation of woodland, much of which has been devastated by grazing. The combined effects of the removal of feudal obligations, the Industrial Revolution and a rapid increase in population led to these changes. The landscape was altered significantly by the switch from fallow rotation agriculture to crop rotation agriculture. Significant changes were also brought about by the drainage of moorland, marshes, and periodically wet, heavy soils; this increased the area of available cropland and allowed the intensification of grassland management. These processes have continued through the twentieth century and have led to a gradual loss of biodiversity within the landscape, and to increasing environmental damage, particularly from groundwater pollution with nitrogen and pesticides, soil erosion and contamination of surface waters with phosphates. Mediterranean areas are increasingly affected by the process of agricultural industrialisation. Cereal production is spreading and production of citrus (and other) fruits, wine and vegetables has expanded and intensified. Traditional forms of cultivation have, accordingly, been put under pressure. Since grazing in mountain areas continues, the revival of the forest has, however, been slowed.

Agricultural development in most of the former communist countries has been characterised by several periods of sudden change. Large estates were broken up and huge agricultural enterprises created through collectivisation. This process also saw the growth of central villages (while smaller villages stagnated) and the loss of family farms. In former East Germany the average size of an arable farm in 1989 was about 5000 ha. Livestock enterprises were run as independent units and, on average, about three such enterprises would

6

be found somewhere on the land occupied by an arable farm. This concentration of cattle and pig husbandry made slurry disposal very difficult and, in some areas, resulted in excessive nitrogen application to the land. The agricultural landscape was dominated by huge arable fields. Within about 2 years of the democratic 'revolutions' (i.e. by 1992), the agricultural production cooperatives had been broken up and a totally different farm structure had developed. Interestingly, there was no return to the pre-communist structure, where family farms of between 15 and 25 ha dominated, complemented by estate farms of up to 100 ha and smaller farms of less than 15 ha. Large agricultural enterprises of varying ownership now dominate and the new family farms each manage an average of around 100 ha. Most of the farmland is, however, rented. A small amount of land is farmed by part-timers (see Table 1.1).

The agricultural sector in former East Germany is dominated by arable enterprises. These adjust their production schedules annually in order to account for changes in the agricultural marketplace. The proportion of land put down to fallow follows the latest EU guidelines and changes almost yearly. However, oilseeds for industrial markets are commonly planted on the fertile soils, while the large areas of fallow are found on poorer soils. The agricultural landscape is still characterised by large arable fields and the widespread absence of small biotopes. In the two years following 1989, the cattle population dropped by 50%, and the pig population by two-thirds. The application of animal manure has declined massively, with a concurrent reduction in the nitrogen burden on soils. In many areas, this has also reduced the risk of groundwater contamination with NO_3. The number of people employed in agriculture has plummeted (Heinrich, 1994; Krönert, 1994b).

The restructuring of agriculture has, however, taken a different course in each country. In Romania, for example, privatisation has produced a farming structure dominated by small and very small enterprises. In Russia the transition to new enterprise forms is progressing very slowly. Mander *et al.*

Table 1.1 Farm structure in Saxony-Anhalt

Legal status	1992			1995			
	Number of farms	% of farms	% of agric. area	Number of farms	% of farms	% of agric. area	Av. farm size (ha)
One-man farms	2764	74.0	20.5	3224	72.22	23.75	
Full-time	1591	42.6	17.7	1787	40.03	21.09	137.0
Part-time	1173	31.4	2.8	1437	32.19	2.66	21.5
Partnerships	416	11.1	15.0	680	15.23	22.78	388.9
Business corporations	217	5.8	17.7	225	5.14	15.76	813.2
Registered coops	335	9.0	46.7	295	6.51	37.27	1464.4
Others	5	0.1	0.0	40	0.9	0.5	145.1

Source: Heinrich, 1994; Heinrich *et al.*, 1996

(1994) has described the society-driven changes to agricultural structures in Estonia and the resulting impacts on the agricultural landscape.

The original approach of the MAB project – that the investigation of regional differences in the current condition and development pattern of agricultural regions and landscapes in each part of Europe must be tackled on an individual basis – proved most appropriate. It became very clear that sudden changes in the development of individual agricultural areas are by no means impossible, and that indicators of such changes have to be taken very seriously if constructive intervention is required.

THE ROLE OF THE AGRICULTURAL ENTERPRISE

Baudry and co-workers (Baudry and Denis, 1992; Baudry *et al.*, 1992a,b; Denis *et al.*, 1992) were prominent in the development of the generally accepted hypothesis that the farm enterprise represents the binding element between the socioeconomic or political conditions in which agriculture operates and the local agroecological system. They underpinned this hypothesis with examples from Normandy (Pays d'Auge). Accordingly, farm mosaics, land-use mosaics and ecological mosaics need to be considered when studying real landscapes (Denis *et al.*, 1992). A holistic approach which takes into account the theory of system hierarchies has proved itself to be most appropriate when studying land-use systems in a region. The holistic and system approach allows the subject area to be split into system parts. Each part can then be studied individually and the relationships between the parts explored. In the study of land-use systems, the land-use mosaic is taken to be a subset of the farm. The division into system parts allows the relationships between farm type, land use and ecosystem to be better understood. The Pays d'Auge study, for example, showed that social factors are currently more important determinants of land use than natural factors. It was found that the so-called marginal farmers – farming just 20% of the land – are particularly important for the preservation of biodiversity (Baudry and Denis, 1992). The region was dominated by permanent grassland, but with a recognisable trend towards arable production. It is believed that the new Common Agricultural Policy (CAP) will cause land-use change, with inevitable ecological consequences. The effects of the CAP on landscapes are not straightforward, however, since they are mitigated by farming systems. In reality, it is farmers' decisions that change land use.

The linking of ecological and socioeconomic approaches and the transition from a lower hierarchical level to the next highest proved to be a difficult methodological problem. Baudry *et al.* (1992) started from the basic idea that landscape structure is determined by land-use/land cover where the field is considered the smallest unit of relevance. Another basic idea is that the field is bound up within both a socioeconomic and a landscape hierarchy. Given this, then there exists a socioeconomic hierarchy (field → agricultural enterprise → local community → socioeconomic region) and a landscape hierarchy (field → land-use mosaic → group of land-use mosaics of various spatial dimensions). A combined, comparative study at the middle hierarchical level is then made possible by relating statistical data about a community to a set of grid-based land-use data.

Hierarchical approaches have also been considered in landscape evaluation. Diemann and Bauer (1994) discuss case studies in the black-earth soil region west of Halle formerly occupied by two large agricultural enterprises. The intention is to determine the ecological value and economic profitability of each individual field, and then to aggregate the results to derive net outcomes for individual enterprises and the region as a whole.

Milanova *et al.* (1994a,b) also work with several hierarchical levels. An individual enterprise (at Borodino, west of Moscow) represents one observation level which is then subdivided according to the type and intensity of land use on that enterprise. Milanova *et al.* identified ten broad land-use types and these differ in terms of the proportion of arable land or permanent grassland, and dosage and frequency of fertiliser/manure application, and the intensity of pesticide use.

Discussion of the importance of the agricultural enterprise has helped in understanding that farm enterprises are bound up with a hierarchy of socio-economic regions, and that the enterprise's operational units (the fields) are bound up within a hierarchy of landscapes.

In addition to the complex studies of landscapes and agricultural regions, part of the MAB project was also devoted to special issues concerning the relationships between man and the environment (in the context of agricultural landscapes). Investigation of biological and landscape diversity and the flow of material within agricultural landscapes is worth further mention.

BIOLOGICAL AND LANDSCAPE DIVERSITY IN AGRICULTURE

Burel (1994) has made a number of basic observations concerning biotope networks and biodiversity in the landscape. The movement of organisms in a landscape network does not follow a common pattern. The movement of some species' populations is enhanced by linear landscape elements that act as corridors, while that of others is stopped by the same elements acting as barriers. Some organisms react at such a scale that they are unable to perceive these landscape elements, either because the organisms are too small and do not move, or because they are highly mobile. There is no universal relationship between landscape structure and the movement of organisms. The relationship depends on a range of biological characteristics of the population in question. These characteristics include size of home and range, dispersal mechanisms and capacities, and sensitivity to fragmentation. Even if the populations being studied use corridors, the efficiency with which a corridor supports the movement of the organisms can vary. Among the important determinants of this efficiency are vegetation structure (herbaceous, shrub and tree layers), corridor width, edge structure, and plant species composition. The presence of a corridor, or even of connected corridors, does not therefore necessarily ensure the dispersal of a population. This may be because of poor corridor quality or low species mobility. Even if corridors are efficient, they may be harmful for a species population; individuals may concentrate on this route and attract predators. Last but not least, corridors may also enhance the movement of pests or diseases across a landscape. Assuming corridors do play an important role in the maintenance of species diversity at a regional level we have yet to define which species, time-scales and structural conditions this applies to. More field research and modelling are needed so that we can provide more detailed advice to planners and managers on how to design landscape networks which support the survival of wildlife in agricultural landscapes. These more detailed aspects are being investigated for beetles in the hedge landscapes of Brittany, France. It is also interesting to note that the dynamics of species' patterns follow those of landscape patterns, but are separated by an interval of time.

These findings make it clear that any studies of an (indicator) species have to be carried out on a one-by-one basis. Such issues were rarely looked at in the MAB project. There can be no doubt, however, that hedges are of general and paramount importance to the diversity and abundance of birds. This has been clearly demonstrated by Kujawa (1994) in Poland. A comparative analysis of the breeding bird communities of five study areas covering various types of agricultural landscapes indicated that land use (particularly hedge structure and the proportion of land taken up by small fields) has considerable influence on the abundance and diversity of breeding birds. Hedges play a decisive role in maintaining populations of breeding avifauna. In the Polish research up to 60% of all breeding pairs of birds nested in the structurally rich hedges, even though the latter only occupied between 2 and 4.2% of the total land area investigated. The highest combined nesting densities of typical field species – such as the corn bunting (*Emberiza calandra*), yellow wagtail (*Motacilla flava*), partridge (*Perdix perdix*) and quail (*Coturnix coturnix*) – were found in those areas with a greater proportion of small fields. Using the example of the western curlew (*Numenius arquata*), Vogel and Slotta-Bachmayr (1994) were able to show that specific measures for the protection of individual endangered species are successful where:

(1) There is detailed knowledge of the habitats of these species, in terms of size, spatial distribution and plant species composition; and
(2) measures have been taken to recreate these habitats.

A further requirement for successful protection of endangered species is the acceptance of protection measures by local land users. These users must also be compensated financially for any 'losses' incurred through these measures, for example if there are restrictions on mowing frequencies on grassland.

MATERIAL FLOWS IN THE AGRICULTURAL LANDSCAPE

The Ryszkowski-inspired inclusion of flows of material within the agricultural landscape in the MAB working group discussions is of fundamental importance. Agriculture is the main source of continuous environmental damage. This damage is caused by wind and water erosion and by contamination of groundwater and surface waters with nitrogen, phosphorus and pesticides. Measures for reducing soil erosion would help to solve these environmental problems. The use of fertilisers and pesticides also has to be reduced so that pollution levels drop below critical threshold limits. Shelterbelts and wide strips of meadow located alongside ditches or streams and in dells or v-shaped valleys can protect water from nutrient contamination. The survival and spread of woody perennials and meadows in the agricultural landscape tends to be seen exclusively in terms of an improvement in biodiversity; endangered animal and plant species have a better chance of surviving if provided with connections or stepping stones between separate habitat patches. Further investigation into the interactive effects of shelterbelts, field trees, shrubs and permanent grassland, both in terms of improving biodiversity and manipulating flows of material, is recommended. The Poznań seminar provided considerable stimulation in this direction and some widely relevant issues concerning material flows in the landscape were presented and discussed.

The influence of vegetation cover on water cycling was investigated in Wielkopolska (Poland). The provision of shelterbelts covering 12% of the watershed led to decreased energy use for evapotranspiration in the zone between shelterbelts, evapotranspiration itself reducing from 510 mm to 460 mm. During the plant growth season, shelterbelts themselves have higher evapotranspiration rates than meadows or cultivated fields. In winter and early spring a landscape with shelterbelts can store about 20–60 mm more water in its soils than can an area consisting of open and uniformly cultivated fields.

In areas where potential evaporation is greater than or equal to precipitation, actual annual evapotranspiration is closely dependent on land use, as the values obtained in Wielkopolska show:

Shelterbelt	609 mm
Meadow	500 mm
Oilseed rape field	465 mm
Beet field	454 mm
Wheat field	436 mm
Bare soil	364 mm

Source: Ryszkowski, 1994; Kędziora, 1994.

In controlling the spread of pollutants, shelterbelts and meadows function as biogeochemical barriers. When cultivated fields directly adjoined a drainage channel, the annual leaching of phosphate ions was 33 to 90 mg P/m^2. When the cultivated fields were separated from the drainage canal by an 80 to 90 m wide meadow, the annual leaching of phosphates came to only 12 to 37 mg

P/m² (Ryszkowski, 1994). Ponds and small mid-field reservoirs are also bio-geochemical barriers and are able to prevent the spread of chemical compounds.

Mander and Mauring (1994) use numerous sources to prove that woodland strips and grassland retain nitrogen and phosphorus. They also show that nitrogen retention is more effective in wetlands. However, they point out that the green material produced by these landscape elements has to be removed if the shelterbelts and grassland are not themselves to become sources of pollution.

Rasmussen *et al*. (1994) found that the release of nitrogen from different small watersheds in Denmark varied considerably over the course of a hydrological year. Nitrogen levels were very high in areas with drained clay soils, and considerably lower in sandy areas. Levels of release were lower in watersheds dominated by grassland than in watersheds with a higher proportion of cereal fields. Nitrogen release was also higher in areas with relatively more pig farms. It was also shown that 75% of nitrogen release occurred in autumn and winter. The results of this study indicate that strategies for reducing nitrogen leaching must be based on an understanding of the current hydrogeological and soil conditions within the watershed; these parameters have proved to be more important than the actual crop pattern. Although the crop pattern and livestock situation are still important, it seems that the prevailing (relatively) stable physical conditions are the key factors governing inorganic nitrogen leaching. This implies that detailed investigations of soil water movements and chemical processes in the soil are most likely to improve our understanding of the processes leading to environmental degradation.

A first step toward identifying potential agricultural sources of environmental pollution would be a classification of agricultural enterprises according to their qualitative and quantitative use of fertiliser and manure. Quantification of the flows of materials within the landscape must consider both the naturally occurring material flows and the inputs from agriculture.

DEVELOPMENT PROBLEMS IN THE UKRAINIAN AGRICULTURAL LANDSCAPE

Scientists in the former Soviet Union traditionally took a top-down approach when investigating the dynamic relationship between nature and society. This attitude still prevails today. Studies focus on whole countries or large regions, with some complementary investigations of smaller representative areas. This methodology is reflected in the Ukrainian contributions to the MAB project.

There is a very clear holistic approach behind the work contributed by Voloshyn and Bazilevich (1992, 1994) and other Ukrainian authors (Voloshin *et al.*, 1992a,b,c; 1994). The environmental impacts of agriculture and industry are taken together in an interactive context. Land use is seen as an indicator of the degree to which the local environment is exploited and influenced. This allows categorisation into landscapes for which different development strategies have to be found. In the MAB work the Ukraine is divided into five large regions, each distinguished by the degree of anthropogenic change within the region. The most change is found in the grassland and forest steppes. Five zones can then be identified, each requiring different measures for environmental amelioration:

(1) Humid, warm, temperate, mixed forest: water improvement and regulation of soil acidification.
(2) Broad-leaved forest: water improvement, agrotechnical measures against soil erosion, regulation of soil acidification.
(3) Humid, warm forest steppe: measures against soil erosion, soil improvement (through agrotechnical and hydrotechnical measures and through the use of chemicals).
(4) Very warm steppe zone: measures against soil erosion, improvement of irrigation systems in combination with chemical measures.
(5) Very warm, dry steppe zone: measures against drought (irrigation, protective forest strips, agrotechnical measures).

The Ukrainian part of the Carpathians is classified according to 'dominating' and 'accessory' land uses. This allows rural areas to be incorporated within a broader treatment of economic areas, thus putting their importance into proper context. This is made clear in the following overview in Table 1.2.

Many Ukrainian studies deal with the current condition, degradation and loss of arable land. They all reach the conclusion that urgent measures are required to prevent soil erosion and the loss of soil organic matter. On slopes between 3° and 7°, Tarariko (1992a,b) recommends crop rotations which ameliorate erosion, with a grassland proportion of 30–40%. Steeper slopes should be afforested or converted to permanent grassland.

Until the beginning of the 1990s, Ukrainian agriculture was dominated by large enterprises with huge arable fields. The importance of different enterprise structures have not, therefore, been analysed. However, the Ukrainian

Table 1.2 'Dominating' and 'accessory' land-use forms in the Carpathian economic region of the Ukraine

Dominating land use	Accessory land use
Industrial belt with chemical industries and oil refineries	Intensive agriculture and forestry
Area with coal-fired power stations, chemical industry etc.	Intensive agriculture and forestry
Belt of intensive agriculture with a high proportion of arable land	Local industrial centres
Belt of extensive agriculture with a low proportion of arable land	Forestry, recreation
Forestry	Recreation, timber industry

Source: Voloshyn and Bazilievich 1992, 1994

MAB work highlights the widespread and catastrophic soil degradation caused by large-scale agriculture and suggests measures for avoiding such degradation. 'Biological amelioration' through afforestation, grassing up and organic enrichment are the most promising options. Given the circumstances in the Ukraine, there is an urgent need to improve biodiversity, especially since this will also contribute to the prevention of soil erosion and the control of material flows in the agricultural landscape.

THE PORTUGUESE EXAMPLE

Since one of the aims of the MAB project was to carry out regional comparisons, the potential for applying those land-use analysis methods used in NW Europe to Portugal was investigated (Pinto-Correia, 1992; Reenberg and Pinto-Correia, 1994). Such a method transfer only proved possible to a limited extent. Rural areas in NW and central Europe are made up of clearly delineated fields and meadows, landscape patches (e.g. woods, ponds, lakes, settlements) and linear landscape elements (e.g. rivers, ditches, paths, streets, grassland strips). In contrast, the Mediterranean landscape features a huge range of transitional land-use states. The authors illustrate this point using the montado landscape – originally an agrosilvopastoral land-use system. This system has been transformed into very different types of land use, ranging from intensive arable cropping through to *Eucalyptus* plantations, but also including abandoned fields covered in secondary vegetation. The authors use the identification of land-use types, distinguished using dominant land use and the land-use mosaic, as a means of addressing landscape characterisation. Further characterisation is then achieved by distinguishing between different types of vegetation cover and by using faunal characteristics. The importance and potential development trends of each landscape type are also examined. This approach is indispensable if some way of understanding the landscape is to be found. It also represents an alternative which could be applied to the study of land-use structures in central and NW Europe.

METHODS FOR LANDSCAPE ANALYSIS AND DIAGNOSIS

Rural areas and agricultural landscapes are very complex, as are the biodiversity and material flows within them. This complexity has forced us to select methods from a range of perspectives when undertaking landscape analysis, diagnosis, prediction and planning. Bastian (Chapter 3) summarises the approach of the Leipzig–Dresden School of Landscape Ecology, where the key is an understanding of the natural complex and its potential exploitation. Taking the perspective of the Munich School of Landscape Ecology, Duhme *et al.,* (1994) have presented another methodological approach for use when planning a comprehensive environmental protection programme. First, the authors define the objective of an evaluative process and the landscape functions desired for the area under consideration. They make concrete suggestions for a 140 km² area of land near Munich (Germany). These include the establishment of a forest protection system, reduction of soil erosion (with an implicit reduction in water pollution), and improvement in the conditions for native game species. These objectives are then defined numerically; if the objectives are to be achieved, then the wooded area should be 14% of the total area, soil erosion should be less than 7 tons per hectare per year, and there should be 100 m of hedgerow per hectare of agricultural land. The current condition of the landscape is recorded on analytical charts and each individual landscape unit (each unit is an average area of 3–4 km²) is then evaluated in terms of the objectives. The result of the evaluation is presented graphically on maps containing the recommended forest protection system, any deficit in linear structures used for controlling soil erosion, and any deficit in the hedgerow system. To achieve the required objectives in the sample area, for example, 12.85 km² of arable land would have to be converted to woodland, and 10 km² (on steep slopes or in marsh valleys) converted to extensive grassland. A further 2.68 km² of agricultural land would need to be appropriated for hedgerow planting. In this scenario, tree planting alone would cost DM 103 million (about USD 57 million). It is clear that any transformation of the agricultural landscape to meet the requirements of environmental, soil and water protection would be extremely expensive and could only be possible with financial support from some kind of extensification programme. From a methodological viewpoint, the sequential implementation of procedural steps is worth noting, i.e. analysis of current condition → definition of evaluation objectives and desired landscape functions → quantification of these objectives to give a numerical evaluation system → identification of deficits in the current state of the landscape → suggestion of land-use changes → calculation of the costs of these changes.

APPROACHES TO AGRICULTURAL DEVELOPMENT

Breitschuh *et al.* (1994) have put forward a framework for a land-use approach (EULANU) which would be both economically efficient and environmentally friendly, and which would involve widespread extensification of agricultural production. The plan is based on a multifunctional agricultural enterprise which derives its income not only from producing food but also from bioenergy production and conservation of the cultivated landscape. Their suggestion assumes that the production of bioenergy would mostly take place in those regions less suited to the production of food, and that 1–10 MW biomass-fuelled power stations would be built in these areas. A 20–30% reduction in food production would be expected. In EULANU, both food production and bioenergy production should be environmentally friendly. Critical threshold values would be defined for 32 measures of environmental damage; the actual values set would vary between regions. In theory, these critical limits should not then be exceeded and an appropriate monitoring system would need to be established. Examples of the kinds of measures suggested are annual net nitrogen surplus, acceptable amount of soil erosion per hectare per year, maximum amount of annual pesticide application, and livestock stocking densities. In addition, farmers would be obliged to undertake measures contributing to positive countryside management. They would be paid for doing so, provided the critical threshold values mentioned earlier are not exceeded on their land. Both machinery and livestock (sheep and cows) would be used in landscape management. To a small extent, agricultural land would also be converted into biotopes in order to raise biodiversity. Some sample calculations in Thüringen (Germany) have shown that the net costs of this approach would be no higher than those incurred through current agricultural support measures. The advantage of this new approach is that agricultural products could be sold at world market prices, while at the same time environmental conditions could be improved considerably (e.g. through carbon-neutral energy production and reduction of water pollution from nitrates, phosphates and pesticides). Interestingly, these suggestions not only originated from within the agricultural sector itself, but they also seek to preserve social structures in rural areas while providing financial compensation for particular landscape management activities. It should, however, not be forgotten that the costs of a permanent countryside monitoring system would be considerable.

The basic principles involved in planning for biodiversity were already formulated as early as 1992, by Merriam, a Canadian guest. An understanding of how ecological systems function is a prerequisite for such planning. In order to understand species loss and to be able to control biodiversity we must know enough about relevant ecological processes. Another prerequisite is an understanding of the effects of agriculture on biodiversity. Farming is profit-orientated, but society increasingly expects farmland to incorporate and protect natural features whose social values are now high and increasing. It is argued that these natural features are themselves of long-term value to the profit-making farming operation itself.

Due at least in part to such ethical and social considerations, regional land-use planning (where applied to farmland) commonly addresses issues at the interface between land-use regulation and conservation planning. This planning often includes consideration of the biodiversity of farmland regions.

The simplest way of improving biodiversity in agricultural areas is to stop using the land for agriculture, thereby increasing species diversity. Some kind of moderate interference is still required, however, in order to ensure the continuation of this diversity. A second approach is the establishment of so-called beta-diversity. This involves raising diversity along a transect, at the same time allowing different habitat development stages to exist (for example, in woodland parcels). The time component needs to be considered in such planning approaches: colonisation has to occur naturally. The challenge is then to maintain a species presence in the region long enough to allow recolonisation to occur.

A key component of biodiversity management is the maintenance of landscape connectivity so that barrier effects are reduced. Examples of connective landscape elements are riparian strips, hedges, field margins, livestock trails, and tunnels and bridges under or over roads. The reduction in net production intensity as well as the complete abandonment of agriculture should also be considered when managing for biodiversity.

LANDSCAPE MONITORING

Given that changes have taken place in agricultural regions, and that further significant changes in European agricultural landscapes are very likely, landscape monitoring is of increasing importance. This monitoring can be applied to individual landscape elements. Brandt (1992) has described one such approach involving the classification and mapping of small biotopes (small uncultivated areas that serve as habitats for wildlife within the agricultural landscape) in Denmark. He divides these small biotopes into line biotopes and patch biotopes. These are then further subdivided into wet and dry biotopes and finally into individual biotope types. Thanks to changes in market conditions and environmental policy, dramatic changes in Danish agricultural land-use patterns have been experienced since the middle of the 1980s. In this respect, small biotopes are a useful indicator of the degree of land-use intensification in the surrounding agricultural region. Differences in the pattern of small biotopes were therefore analysed through time in a number of test areas, and related to agricultural structural and technological developments. In Denmark it seems that the number of small biotopes has stabilised following a long period of decline (which particularly affected wet biotopes), and is even beginning to increase.

Systematic observations of large-scale land-use changes in rural areas have been made possible with the arrival of satellite remote sensing. The loss of agricultural land to urbanisation or afforestation can also be tracked using this technique. Folving and Megier (1992) discuss potential applications of remote sensing in their report on the Collaborative Programme on Application of Remote Sensing in the Management of the Less Favoured Areas. The programme was managed by the Joint Research Centre of the European Commission in Ispra (Italy). They recommend creating a digital ecological map of Europe as a complement to the CORINE Land Cover Map. Satellite remote sensing has already found wide application since land-use changes are fairly easy to identify when records from different years are compared. Remote sensing is an excellent tool for identifying regions of dynamic activity and regions with rapidly changing land uses should be monitored on a systematic basis. The technique is, however, unable to identify the cause of such changes, or to predict the ecological consequences – terrestrial methods of landscape analysis and diagnosis are still needed.

REFERENCES

Baudry, J. and Denis, D. (1992). Land-use systems in the Pays d'Auge. In Baudry, J., Burel, F. and Hawrylenko, V. (eds.) *Comparisons of landscape pattern dynamics in European rural areas*. 1991 seminars, EUROMAB Research Programme, Lieury, Rennes, Kiev, pp. 139–50

Baudry, J., Burel, F. and Hawrylenko, V. (eds) (1992). *Comparisons of landscape pattern dynamics in European rural areas*. 1991 seminars, EUROMAB Research Programm. Lieury, Rennes and Kiev, 355 pp.

Baudry, J., Denis, D., Lecroq, V., Laurent, C. and Vivier, M. (1992a). Le Pays d'Auge, a region and a landscape. In Baudry, J., Burel, F. and Hawrylenko, V. (eds). *Comparisons of landscape pattern dynamics in European rural areas*. 1991 seminars, EUROMAB Research Programme, Lieury, Rennes, Kiev, pp. 97–103

Baudry, J., Laurent, C. and Denis, D. (1992b). A hierarchical framework for studying land cover pattern changes from an ecological and economical stand point: concepts and results in the Normandy research. In Baudry, J., Burel, F. and Hawrylenko, V. (eds). *Comparisons of landscape pattern dynamics in European rural areas*. 1991 seminars, EUROMAB Research Programme, Lieury, Rennes, Kiev, pp. 104–14

Bauer, S. (1994). EC agricultural policy and its impact on land use and environment. In Krönert, R. (ed.) *Analysis of landscape dynamics – driving factors related to different scales*. 1993 symposium, EUROMAB Research Programme, Leipzig, pp. 19–29

Bowler, I. and Ilbery, B. (1992). A conceptual framework for researching the environmental impacts of agricultural industrialization. In Baudry, J., Burel, F. and Hawrylenko, V. (eds). *Comparisons of landscape pattern dynamics in European rural areas*. 1991 seminars, EUROMAB Research Programme, Lieury, Rennes, Kiev, pp. 288–97

Brandt, J. (1992). Land use, landscape structure and the dynamics of habitat networks in Danish agricultural landscapes. In Baudry, J., Burel, F. and Hawrylenko, V. (eds). *Comparisons of landscape pattern dynamics in European rural areas*. 1991 seminars, EUROMAB Research Programme, Lieury, Rennes, Kiev, pp. 213–29

Breitschuh, G., Eckert, H. and Ahl, C. (1994). Efficient and environmentally sound land use (EULANU) – blueprint for farm management and landscape conservation on market economy lines. In Krönert, R. (ed.) *Analysis of landscape dynamics – driving factors related to different scales*. 1993 symposium, EUROMAB Research Programme, Leipzig, pp. 51–9

Burel, F. (1994). Biological fluxes in landscape networks: role in the maintenance of regional biodiversity. In Ryszkowski, L. and Bałazy, S. (eds). *Functional appraisal of agricultural landscape in Europe*. 1992 seminars, EUROMAB Research Programme and INTECOL, Poznań, pp. 177–82

Denis, D., Baudry, J. and Al Jallad, A. (1992). Technical note: data management principle to study farming systems and landscapes at several scales. In Baudry, J., Burel, F. and Hawrylenko, V. (eds). *Comparisons of landscape pattern dynamics in European rural areas*. 1991 seminars, EUROMAB Research Programme, Lieury, Rennes, Kiev, pp. 115–22

Diemann, R. and Bauer, S. (1994). Development of land use in rural areas in East Germany since 1989. In Krönert, R. (ed.) *Analysis of landscape dynamics – driving factors related to different scales.* 1993 symposium, EUROMAB Research Programme, Leipzig, pp. 42–50

Duhme, F., Pauleit, S., Schild, J. and Stary, R. (1994). Quantifying development targets for nature conservation in rural landscapes. In Ryszkowski, L. and Bałazy, S. (eds). *Functional appraisal of agricultural landscape in Europe.* 1992 seminars, EUROMAB Research Programme and INTECOL, Poznań, pp. 249–66

Folving, S. and Megier, J. (1992). Remote sensing applied in management of less favoured areas and landscape ecological mapping experiences from the JRC European collaborative programme. In Baudry, J., Burel, F. and Hawrylenko, V. (eds). *Comparisons of landscape pattern dynamics in European rural areas.* 1991 seminars, EUROMAB Research Programme, Lieury, Rennes, Kiev, pp. 308–18

Heinrich, J. (1994). Changes of farm structure and land use in Saxony–Anhalt. In Krönert, R. (ed.) *Analysis of landscape dynamics – driving factors related to different scales.* 1993 symposium, EUROMAB Research Programme, Leipzig, pp. 30–41

Heinrich, J., Gersonde, J., Hüwe, R. and Wiesner, F. (1996). Stand der Umstrukturierungen der Landwirtschaft im Land Sachsen–Anhalt. In Institut für Agrarökonomie und Agrarumgestaltung (eds). 6. *Mitteilung. Landwirtschaftliche Fakultät der Universität Halle-Wittenberg,* Halle

Kędziora, A. (1994). Energy and water fluxes in an agricultural landscape. In Ryszkowski, L. and Bałazy, S. (eds). *Functional appraisal of agricultural landscape in Europe.* 1992 seminars, EUROMAB Research Programme and INTECOL, Poznań, pp. 61–76

Krönert, R. (ed.) (1994a). *Analysis of Landscape Dynamics – Driving Factors Related to Different Scales.* 1993 seminar, EUROMAB Research Programme, Leipzig, 164 pp.

Krönert, R. (1994b). Recent changes of land use and their driving factors in the rural area of the Leipzig region (Saxonia) – ecological consequences of land use changes. In Krönert, R. (ed.) *Analysis of landscape dynamics – driving factors related to different scales.* 1993 symposium, EUROMAB Research Programme, Leipzig, pp. 60–72

Kujawa, K. (1994). Influence of land-use change within agricultural landscapes on the abundance and diversity of breeding bird communities. In Ryszkowski, L. and Bałazy, S. (eds). *Functional appraisal of agricultural landscape in Europe.* 1992 seminars, EUROMAB Research Programme and INTECOL, Poznań, pp. 183–96

Mander, U. and Mauring, T. (1994). Nitrogen and phosphorus retention in natural ecosystems. In Ryszkowski, L. and Bałazy, S. (eds). *Functional appraisal of agricultural landscape in Europe.* 1992 seminars, EUROMAB Research Programme and INTECOL, Poznań, pp. 77–94

Mander, Ü., Palang, H. and Tammiksaar, E. (1994). Landscape changes in Estonia during the 20th century. In Krönert, R. (ed.) *Analysis of landscape dynamics – driving factors related to different scales.* 1993 symposium, EUROMAB Research Programme, Leipzig, pp. 73–97

Merriam, G. (1992). Biodiversity in temperate agricultural landscapes. In Baudry, J., Burel, F. and Hawrylenko, V. (eds). *Comparisons of landscape pattern dynamics in European rural areas.* 1991 seminars, EUROMAB Research Programme, Lieury, Rennes, Kiev, pp. 319–27

Messerli, B. and Messerli, P. (1978). Touristische Entwicklung im inneralpinen Raum. Konsequenzen, Probleme, Alternativen, *MAB Mitteilunge*, **4**, Bonn

Milanova, E.V., Solntsev, W. M., Kalitskova, N. N. and Gorbunova, L. I. (1994a). Fixes in periurban landscapes of the Borodino collective farm in the Moscow Region. In Ryszkowski, L. and Bałazy, S. (eds). *Functional appraisal of agricultural landscape in Europe*. 1992 seminars, EUROMAB Research Programme and INTECOL, Poznań, pp. 273–82

Milanova, E. V., Tcherkashin, P. A. and Kalutskova, N. N. (1994b). Assessment of landscape suitability for agroeconomic activity (GIS methodology). In Krönert, R. (ed.) *Analysis of landscape dynamics – driving factors related to different scales*. 1993 symposium, EUROMAB Research Programme, Leipzig, pp. 117–31

Pinto Correia, T. (1992). Analysis of the rural landscape, changing perspectives from northern to southern Europe. In Baudry, J., Burel, F. and Hawrylenko, V. (eds). *Comparisons of landscape pattern dynamics in European rural areas*. 1991 seminars, EUROMAB Research Programme, Lieury, Rennes, Kiev, pp. 185–201

Rasmussen, M., Reenberg, A. and Bartholdy, J. (1994). Nitrogen fluxes from agricultural landscapes – comparisons at watershed level. In Ryszkowski, L. and Bałazy, S. (eds). *Functional appraisal of agricultural landscape in Europe*. 1992 seminars, EUROMAB Research Programme and INTECOL, Poznań, pp. 19–30

Reenberg, A. (1992). Guidelines for comparative research on land-use changes and their environmental impact in rural areas in Europe. In Baudry, J., Burel, F. and Hawrylenko, V. (eds). *Comparisons of landscape pattern dynamics in European rural areas*. 1991 seminars, EUROMAB Research Programme, Lieury, Rennes, Kiev, pp. 8–19

Reenberg, A. and Pinto-Correia, T. (1994). Rural landscape marginalization – Can general concepts, models and analytical scales be applied throughout Europe? In Krönert, R. (ed.) *Analysis of landscape dynamics – driving factors related to different scales*. 1993 symposium, EUROMAB Research Programme, Leipzig, pp. 3–18

Ryszkowski, L. (1994). Strategy for increasing countryside resistance to environment threats. In Ryszkowski, L. and Bałazy, S. (eds). *Functional appraisal of agricultural landscape in Europe*. 1992 seminars, EUROMAB Research Programme and INTECOL, Poznań, pp. 9–18

Ryszkowski, L. and Bałazy, S. (eds) (1994). *Functional Appraisal of Agricultural Landscape in Europe*. 1992 seminar, EUROMAB and INTECOL, Poznań, 307 pp.

Ryszkowski, L. and Kędziora, A. (1992). Ecological guidelines for management of agricultural landscape. In Baudry, J., Burel, F. and Hawrylenko, V. (eds). *Comparisons of landscape pattern dynamics in European rural areas*. 1991 seminars, EUROMAB Research Programme, Lieury, Rennes, Kiev, pp. 268–71

Tarariko, O. (1992a). Optimization of agricultural landscapes in the Ukraine. In Baudry, J., Burel, F. and Hawrylenko, V. (eds). *Comparisons of landscape pattern dynamics in European rural areas*. 1991 seminars, EUROMAB Research Programme, Lieury, Rennes, Kiev, pp. 57–8

Tarariko, O. (1992b). Optimization of the rural landscapes in the Ukraine using soil-protecting systems of agriculture. In Baudry, J., Burel, F. and Hawrylenko, V. (eds). *Comparisons of landscape pattern dynamics in European rural areas*. 1991 seminars, EUROMAB Research Programme, Lieury, Rennes, Kiev, pp. 59–62 van Latestejn, H. C. (1994). Ground for choice: a policy-oriented survey of land-use changes in the E. C. In Ryszkowski, L. and Bałazy, S. (eds). *Functional appraisal of agricultural landscape in Europe.* 1992 seminars, EUROMAB Research Programme and INTECOL, Poznań, pp. 289–98

Vogel, M. and Slota-Bachmayr, L. (1994). Effect of a nature conservation programme on the meadow breeding bird community. In Ryszkowski, L. and Bałazy, S. (eds). *Functional appraisal of agricultural landscape in Europe.* 1992 seminars, EUROMAB Research Programme and INTECOL, Poznań, pp. 205–24

Voloshin, V. and Bazilewicz, O. (1992). Changes in land use, environment and development patterns for the Carpathians economic region of the Ukraine. In Baudry, J., Burel, F. and Hawrylenko, V. (eds). *Comparisons of landscape pattern dynamics in European rural areas.* 1991 seminars, EUROMAB Research Programme, Lieury, Rennes, Kiev, pp. 42–50

Voloshin, V. and Bazilevich, O. (1994). Land-use diversity in the Carpathian region of Europe: holistic geographical approach and prospects for comparative studies. In Ryszkowski, L. and Bałazy, S. (eds) *Functional appraisal of agricultural landscape in Europe.* 1992 seminars, EUROMAB Research Programme and INTECOL, Poznań, pp. 245–8

Voloshin, V., Bazilewicz, O. and Hawrylenko, V. (1992a). Driving factors of land-use changes and ecogeographical structure of the Carpathians economic region of the Ukraine. In Baudry, J., Burel, F. and Hawrylenko, V. (eds). *Comparisons of landscape pattern dynamics in European rural areas.* 1991 seminars, EUROMAB Research Programme, Lieury, Rennes, Kiev, pp. 51–6

Voloshin, V., Livosky, S., Terlo, V., Telitsky, S., Gavrilenko, V. and Dubin, V. (1992b). Land-use structure in the Ukrainian Territory. In Baudry, J., Burel, F. and Hawrylenko, V. (eds). *Comparisons of landscape pattern dynamics in European rural areas.* 1991 seminars, EUROMAB Research Programme, Lieury, Rennes, Kiev, pp. 30–3

Voloshin, V., Shyshchenko, O., Mel'hytchuk Yu and Tkach, I. (1992c). Some aspects of agriculture intensification in the conditions of anthropogenic transformation of the Ukrainian landscape. In Baudry, J., Burel, F. and Hawrylenko, V. (eds). *Comparisons of landscape pattern dynamics in European rural areas.* 1991 seminars, EUROMAB Research Programme, Lieury, Rennes, Kiev, pp. 34–41

Voloshyn, V., Gukalova, I., Lisovsky, S. and Mischenko, N. (1994). Ecologoeconomic features of agricultural land use in the Donetsko–Prydniprovsky region of the Ukraine. In Krönert, R. (ed.) *Analysis of landscape dynamics – driving factors related to different scales.* 1993 symposium, EUROMAB Research Programme, Leipzig, pp. 98–110

CHAPTER 2

LAND-USE AND LANDSCAPE CHANGES – THE CHALLENGE OF COMPARATIVE ANALYSIS OF RURAL AREAS IN EUROPE

A. Reenberg and J. Baudry

INTRODUCTION

European agriculture has experienced radical changes during the last four decades which have resulted in increased production per hectare, production surpluses, an excess of agricultural land, extensification of marginal land and modification of rural landscapes. The latter includes the greater differentiation between intensive and extensive agricultural regions, as well as within regions. Indeed, the increasing dissimilarity between intensively and extensively utilized land can be observed at all scales of analysis.

In an historical perspective, land-use changes are normal, but developments within recent decades deserve specific attention for a number of reasons:

(1)   Agricultural intensification has been based on an increasing input of agri-chemicals and animal manures that pollute soil and water;
(2)   It is broadly recognized that land has to be considered as a limited and non-renewable resource;
(3)   The demand from the urban population for nature and open landscapes for recreational purposes is increasing.

The relative stability and sustainability of land-use systems can be evaluated from two points of view: 1) the reproducible or non-reproducible nature of certain land-use practices; and 2) the environmental consequences of certain land-use practices: reduction in biodiversity (the loss of species and genetic diversity within species), pollution of the watertable, soil erosion, and landscape degradation through the perception of different societal groups, such as the rural, urban and agricultural populations. But both points of view should include the consequences of the spatial arrangement between the cultivated and non-cultivated areas and, for the future, ensure that development trends generated by political and economic factors are environmentally sustainable. The necessary starting point is to understand the interaction between society and nature and how this is determined by cultural and historical parameters.

One approach is through the analysis of *landscape pattern dynamics*: this represents a holistic view of man-made and natural surroundings and attempts to bridge the gap between natural, agricultural, human and urban systems. Within this framework it is possible and relevant to investigate how regulatory tools, at various political or administrative levels, and addressing different parts of the landscape system, influence the environment. Furthermore, the sensitivity of different parts of the landscape to regulatory instruments (local, regional, national, European Community or international) can be investigated. In this way the landscape analytical approach can contribute both conceptual and analytical models for the environmental assessment of the complex 'driving forces' in rural landscape development.

The analysis of landscape patterns, under various natural as well as cultural conditions, can be combined with investigations concerning the correlation between landscape characteristics and environmental health; this approach investigates the environmental consequences to be expected from typical development scenarios in European rural areas. Further, comparative studies of landscape pattern dynamics and their relation to relevant environmental parameters can provide a comprehensive and systematic historical knowledge of environmental and ecological problems, which have followed the changing use of rural areas in the recent past.

The analysis of landscape pattern dynamics and the regional comparison of landscapes, however, is not an easy matter. There are problems concerned with the selection of the spatial scale of analysis, as well as with definitions of common concepts for comparative analysis (Reenberg and Pinto-Correia, 1994). For instance, the scale of analysis must be relevant to the phenomenon of interest, such as species dispersal, aesthetic aspects of landscapes, river pollution, or land transformation.

Recently, a number of groups of researchers have proposed frameworks for the analysis of land-cover and land-use change from different perspectives. For example, Fresco *et al.* (1994), adopting a planning perspective, have sought ways to control changes and implement sustainable land-use systems. The objective of Turner and Meyer (1994), in comparison, has been to model changes and to link them with global climate and biogeochemistry models. Both groups make a plea for a multiscale approach from local (farm, household) to regional and global levels, but neither address their research problems from a landscape pattern perspective. Nevertheless, the spatial arrangement and the dynamics of land use/cover can be considered from this perspective (Reenberg, 1998) and one aim of this chapter is to stress the importance of the landscape approach to rural studies. The argument advanced is that landscape structure itself can be considered as a *'driving force'* for numerous ecological processes (Burel, 1995; Forman, 1995).

COMPARATIVE ANALYSIS OF EUROPEAN LANDSCAPES

The conditions and processes of development in rural areas show genuine differences in various parts of Europe. It is important, therefore, to investigate and compare the 'driving forces' behind the landscape changes in Europe. Here, the availability of data on land use is crucial for appropriate decision making on land-use development (van Duivenbooden, 1995).

It is axiomatic that human activities have heavily influenced most landscapes (Reenberg, 1993). Direct changes have been initiated by the use of the land, for example for primary production (agriculture, forestry, mining), urban growth, infrastructure and nature conservation. Indirect influences include acid rain and global climatic change. The process of change has been ongoing for millennia (Berglund, 1991), but within the last hundred years it has accelerated. The agricultural system of northern Europe, for instance, has changed from an almost closed system producing mainly for local consumption and without imports of farm inputs and energy (e.g. fodder, fertilizers, fuel), to an open system relying heavily on such imports and having a large turnover. The development of transportation, by road and rail, over a short period of time created conditions for rapid change and the specialization of regions for agricultural production (Bazin and Larrère, 1983). These developments created a differentiated spatial pattern in the distribution of crops; whereas earlier wheat had been grown almost anywhere, wheat became mainly located on large, fertile plains.

In many less-developed regions in Europe the agricultural system is only at the beginning of the transformation process. Hypotheses on the environmental and socioeconomic consequences of the transformation of agriculture in these regions may be at least partially formulated and tested on the basis of the historical data from more 'advanced' regions. The south-European regions often envisage a development according to the model of intensification as experienced in northern Europe during the last century, even if the actual trend is in fact marginalization. Although within the EU all regions are subordinated to the same agricultural regulations, they have different trends for development. A comparative analysis might thus be relevant.

The overall development pathways in European rural land use can be seen as variations upon two major tendencies – intensification and marginalization (Baudry and Bunce, 1991). However, in relatively fertile regions, landscapes are stable and do not respond quickly to changes from outside, whereas the landscapes in less fertile regions are very sensitive to changes in overall production and market conditions. They will consequently fluctuate greatly with respect to their general land use even within a relatively short time period (Reenberg, 1990).

The trends envisaged during the 1980s have included a growth of the marginalized acreage following increased productivity of the most fertile agricultural land (Skov-og Naturstyrelsen, 1987; Breuning-Madsen *et al.*, 1990; Gilibert, 1993). Although the actual marginalization has been much less than expected, the difference between intensively and extensively used parts of the rural landscapes has been growing (Laurent, 1991). The 1992 reorientation of the EU Common Agricultural Policy introduced, among other measures, a reduction of agricultural production through set-aside – in effect, compulsory fallowing (CEC, 1991; Stryg, 1992). Even if it can be argued that fallow and marginal land are far from being the same thing (Gilibert, 1993), both contribute to an increase of landscape heterogeneity. It underlines the division of agriculture into an intensively cultivated part, with very few nature elements, and an extensive or non-cultivated part, especially if set-aside fallowing becomes permanent. From an ecological point of view, the main difference is that fallow land is managed and can be orientated towards some defined purposes, while abandoned land, by definition, is not managed.

Research is needed regarding these multiple trends in European rural landscape development. This will provide the necessary knowledge about the relative importance of factors driving the development, the sensitivity of the landscape to various changes (e.g. agricultural production conditions), and the correlation between changes in the landscape and the environment. This knowledge might be transferred to guidelines for development strategies – aiming at a trajectory of development which can ensure a more sustainable agriculture than prevailing under the present intensification and concentration of farming activities.

One of the most pressing issues in contemporary European agriculture is the improvement of the environment, at the same time as maintaining both sufficient food production and the rural population. Farming systems are needed that lead to more sustainable production in both socioeconomic and ecological terms. This requires a thorough understanding of ecosystem functioning, not only at the level of the farmer's fields, but also at the landscape and regional levels; but here is raised the problem of the selection of analytical scales, in time as well as space. What is catastrophic when looking at the agroecosystem at a small scale may be part of a sustainable system on a larger spatial scale (Fresco and Kroonenberg, 1992). Conversely, some fine-scale events can become significant when aggregated at the landscape level. This is the case for hedgerow removal: a single hedgerow is usually not a key landscape element by itself, but as removal becomes widespread, problems appear.

Thus, an interdisciplinary and hierarchical approach, which includes parameters describing ecological, technical, legal, administrative, socioeconomic and cultural conditions is needed to answer the following important questions: what are the most important 'driving forces' of landscape change? when, how and where do they cause landscape change?

ANALYSING LAND USE AND LANDSCAPE SYSTEMS

In this section we discuss some problems arising when different scales of investigation and different academic disciplines are integrated in an analytical approach.

First, the complexity of forces determining land use and landscape systems has been widely acknowledged (e.g. Haber, 1990; Young and Solbrig, 1993; Stromph *et al.*, 1994; Reenberg, 1996; van Duivenbooden, 1995) and it is generally accepted that an interdisciplinary approach is needed to analyse human resource management strategies for such complex systems (Turner and

Meyer, 1994). Not all relevant scientific disciplines need be covered in equal detail, but no research activities should be undertaken without the awareness of the interdisciplinary implications, as outlined in Figure 2.1.

Second, it should be acknowledged that the development of more sustainable strategies of resource management can only be successfully implement if existing land use and landscape system functions are properly understood with respect to their objectives, resilience to external changes, etc. This means that an understanding of such system functions should include the already mentioned interdisciplinarity, as well as the analysis of how processes operating at one scale in the hierarchy influence other levels (Allen and Starr, 1981). A combination of spatial scales, using a hierarchical analytical approach, enables the extrapolation of analytical results from one scale to another.

Third, land-use systems have been defined as the combination of specified land uses on a given land unit (Stromph *et al.*, 1994; van Duivenbooden, 1995). Land-use systems comprise both biophysical and socioeconomic subsystems and various biophysical and socioeconomic factors interact at different spatial levels of a land-use system (at field, farm, regional level, etc.) (Stromph *et al.*, 1994; Reenberg, 1996). Land-use systems also reflect the natural resource management strategies carried out at different spatial levels. An agroecosystem, in comparison, is a specific type of land-use system, that is an ecosystem with an agricultural component in its primary or secondary production compartment.

An ideal analytical framework, therefore, is not only interdisciplinary, it also considers spatial levels of analysis as part of a hierarchy. Here, a series of nested databases can be developed to address different data requirements at different spatial levels of analysis. By nesting the data, the information can be used at different scales without violating their integrity. Agroecological concepts can help to identify the scale and the data required to describe the processes (Dumanski *et al.*, 1994). One major problem is that some elements, such as fields, are nested in other functional units of human activities (e.g. farms) and spatial units (landscape patterns), but farms and landscape patterns do not match (Thenail, 1996).

There remain other important analytical problems. For example, land-use and landscape pattern dynamics cannot be investigated without taking into consideration such matters as data combination, the appropriate scale of analysis, generalization between scales, definition and transferability of concepts, etc. Some of these issues are illustrated in the last parts of the chapter.

Spatial aspects of landscape pattern dynamics

Spatial aspects are crucial to the analysis of landscape pattern dynamics. For example, Reenberg and Pinto-Correia (1994) have shown how spatially registered land-use data are superior to aggregate land-use statistics for revealing important developmental trends. The authors analyse landscape

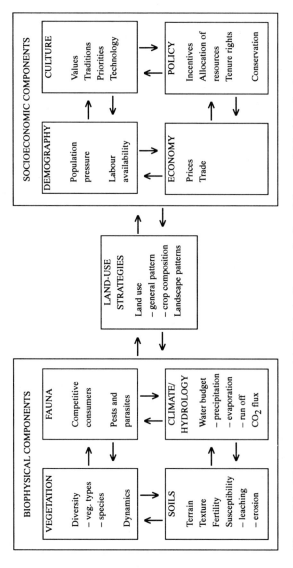

Figure 2.1 Diagrams are often used to illustrate the complexity of forces which drive land-use changes. Modified from Young and Solbrig (1993)

changes in a case study region from a series of topographic maps from 1878 to 1975. A wide range of factors has influenced the variations in land use during the 100-year period (technological capability, socioeconomic conditions, agricultural policies, etc.). In the case presented (Figures 2.2 to 2.6), a radical shift in cultivation intensity from one landscape unit (outwash plains) to another (river valley) can be observed. These developments are not apparent from aggregate land-use statistics (the values shown in Figure 2.7 are the approximate statistics corresponding to the maps), but the trends are important from an environmental point of view.

The lessons learned from the example can be summed up in the following points:

(1) The choice of time series and registration frequency may enhance or hide important characteristics in landscape dynamics.
(2) Quantitative statistical registration of land-use changes may lead to false conclusions, if land use/landscape relations are not included.
(3) The relative importance of forces driving landscape changes varies over time, scale and between landscape units.

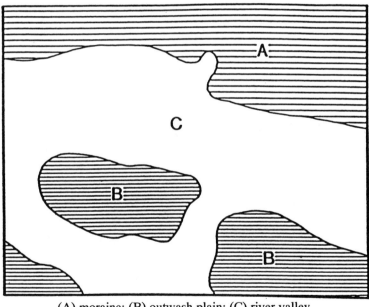

(A) moraine; (B) outwash plain; (C) river valley

Figure 2.2 Geomorphology of the region studied in Figures 2.3 to 2.6, which covers 4 × 3 km. It is located in central Jutland, Denmark. The region has very large variations in natural conditions within short distances. (A) moraine, (B) outwash plain, and (C) river valley (from Reenberg and Pinto-Correia, 1994)

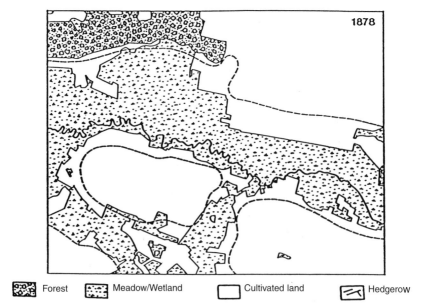

| | Forest | | Meadow/Wetland | | Cultivated land | | Hedgerow |

Figure 2.3 Land use indicated on a topographical map from 1878. The main characteristics were cultivation of outwash plains and extensive use of river valleys. The land-use categories are: forest, meadows, cultivated land, hedgerows (from Reenberg and Pinto-Correia, 1994)

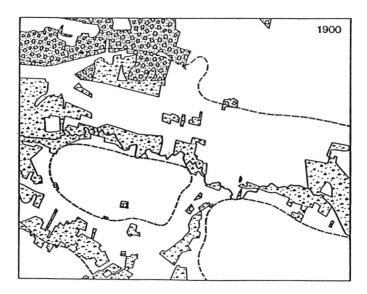

Figure 2.4 Land use indicated on a topographical map from 1900. The main change since Figure 2.3 is the start of intensification in river valleys made possible by new technical developments (e.g. draining). The land-use categories as in Figure 2.3

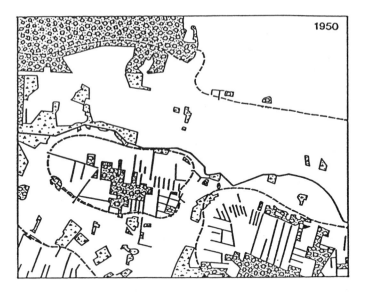

Figure 2.5 Land use indicated on a topographical map from 1950. The main changes since Figure 2.4 are continuous drainage and protection by headgerows on the outwash plains. The categories as in Figure 2.3

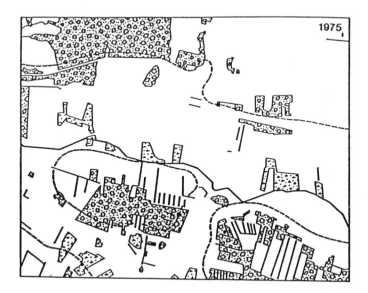

Figure 2.6 Land use indicated on a topographical map from 1975. Cultivation in the river valley is continued and fields on the outwash plains are replaced by plantations. The latter are mainly a result of general changes within agriculture. A shift towards spring-grown cereals in monocultures, as well as the introduction of heavy machinery (combine harvesters) have created conditions under which the sandy soils are too susceptible to wind erosion. The land-use categories as in Figure 2.3

(4) The resilience of landscape varies.

(5) Generalizations about landscape changes across scales in time and space must be handled with care.

Tatoni and Baudry (unpublished data) provide another example of a Mediterranean landscape in Provence (France). In the study area there has been an increase of woodland at the expense of cropland (Figure 2.8). The spatial analysis demonstrates that cropland is not immediately transferred to rough pasture and woodland. It is first replaced by a perennial crop (vineyard or olive grove), which is later abandoned. In the meantime, some woodland is converted into agricultural land. This leads to faster local changes than those perceived on average and it has been shown (Baudry and Tatoni, 1993) that current vegetation is correlated with the trajectories of landscape changes.

The influence of space and time resolution on analytical results

Landscape ecologists have discussed the challenge of incorporating hierarchical and spatial characteristics in the analysis of ecological systems (O'Neill, 1989; Wiens, 1989; Baudry, 1992). The following aspects must be taken into consideration: the scale dependencies of observation methods; data treatment; the functional organization within the hierarchy; and the generalization of research results from one temporal or spatial scale to another. A particular concern is how to estimate the ramifications of local processes on global processes, especially when one cannot easily describe the latter as large-scale variations are involved. One classical example of scale dependence is the number of species: this increases with the size of area studied, so that the co-occurrence of species varies with the extent of the sampled area. Another example is the nitrogen content of water in a watershed. In the buffer zones, denitrification occurs and there is less nitrogen in the river than predicted from leaching rates from individual fields (Haycock and Muscott, 1995).

Using data drawn from statistics to describe the land-use changes in western France for a century, Baudry (1992) has shown that the calculated rates of change are highly dependent on the spatial and time resolution used to measure them. The finer the resolution scale, either for time or space, the faster the apparent changes (Figure 2.9). These results demonstrate that comparisons across landscapes or regions can only be made when similar resolutions of scale are used. Otherwise, a conclusion such as 'changes are more important in region X than in region Z' are meaningless.

Linking different data types

The quantitative integration of biophysical and socioeconomic information is a major problem to be solved in landscape pattern analysis. Land-use strategies are constrained to a certain extent by biophysical conditions, as well as by human decisions. The afforestation of farm land (marginalization) in Denmark

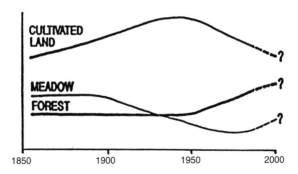

Figure 2.7 The values are approximate land-use statistics for the test region in Figures 2.2–2.6. Land-use statistics do not reveal important trends in development. The shift in cultivation from sandy soils to river valley cannot be seen but is important from an environmental point of view (from Reenberg and Pinto-Correia, 1994)

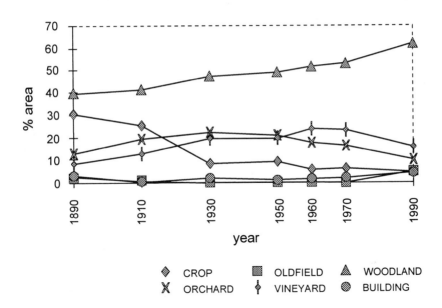

Figure 2.8 Patterns of change in a Mediterranean landscape: the relative area of each land-cover class

provides an illustrative example (Barreusoe, 1987; Reenberg, 1990; Brandt *et al.*, Chapter 5). The broad regional pattern of abandoned farm land matches the pattern of coarse, sandy soils, and this correlation can easily be obtained because of available soil and land-use databases. If we look into the details of the afforestation pattern a very complicated mosaic can be observed; but again all plantings are found on coarse, sandy soils. A reasonable hypothesis, which can explain the patchwork, is that farm-related parameters determine the

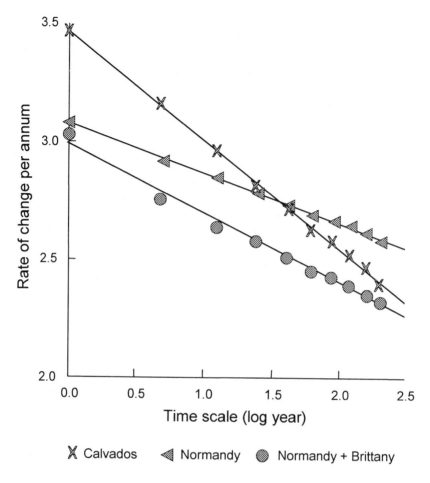

Figure 2.9 Scale dependency in land-use change for three areas of increasing size, from Calvados to Brittany and Normandy. The perceived rate of change increases when the time scale for measurement increases

plantation pattern at a local level. Thus, farm-specific socioeconomic parameters have to be combined with geo-related information on land use and natural resources to provide a suitable analytical model. Socioeconomic and physical factors do not, however, control land-use patterns at a similar scale. The above example shows that the physical environment plays a role at both very coarse (region) and very fine scales (fields), while farmers' decisions control the process at an intermediate scale. Similar results have been obtained by Baudry *et al.* (1993) for the distribution of grassland in Lower Normandy (France).

   To solve this problem, Stromph *et al.* (1994) propose a quantitative method to combine the biophysical and socioeconomic subsystems of a land-use system; nevertheless, important spatial aspects are not explicitly taken care of in their approach. Land-use strategies must also be analysed in a spatial context. First, because the distance factor might pla an important role for land-use decisions. Second, because the spatial configuration of landscape units and natural resource bases might be an important determinant for the environment. Andriesse *et al.* (1994) have suggested an integrated transect method (ITM) to meet this demand. Their approach aims at a systematic description of agroecosystems in a multi-scale framework. It includes biophysical characterization and land-use descriptive parameters such as crop types, cultivation intensity, etc., for a given transect of land. The need for the consideration of demographic, socioeconomic and institutional factors is acknowledged, but has not yet been implemented in their analytical framework.

Methods which combine a land-use description, equivalent to the ITM method referred to above, with a socioeconomic inventory of the households responsible for the land-use decisions are a challenge in landscape pattern analysis. Modern technology, such as Geographical Information Systems, in combination with database systems provide us with promising tools and perspectives (Reenberg and Fog, 1995; Reenberg, 1998).

Transferability and definitions of concepts

The commonly accepted methodology for landscape pattern analysis has been developed for landscapes in intensively farmed agricultural regions in northern Europe. It makes good sense to classify these landscapes into 'matrix', 'patches' and 'corridors' (Forman and Godron, 1986), as they are characterized by well-defined spatial boundaries (Gulinck, 1986). It is also in this type of landscape that the analysis of ecologically relevant characteristics such as 'heterogeneity', 'connectedness' and 'connectivity' has been developed. Thus agricultural fields are considered as a 'matrix', and the structure of the landscape is defined by the 'patches' and 'corridors' of other types of vegetation cover. A landscape pattern can be identified and quantitatively defined, and changes over the time assessed.

This type of classification cannot be directly transferred to all types of landscape: as recognized by Forman and Godron (1986), the 'matrix' element is often difficult to identify, for example in Mediterranean landscapes. Some elements of the classification method may also be less relevant, at least as regards the earlier stages of landscape transformation or the scales (resolutions) suitable for northern European landscapes (Pinto-Correia, 1993; Reenberg and Pinto-Correia, 1994). For example, the absence of a clear pattern in the agroforestry landscape makes it difficult to individualize landscape elements and so define and measure a structure. In such cases it might be better to rely on a classification based on land-use systems (Vos and Stortelder, 1992). Here the boundaries of land-use units can be defined according to parameters such as types and densities of various vegetative elements.

This does not necessarily imply that concepts and methods cannot be transferred, but some further consideration is needed. For example, the suggested land-use systems approach might be successfully applied to a northern European landscape but the relevant scale for observation would probably be at a larger scale of analysis. Likewise, the distinction between 'matrix' and 'patches'/'corridors' might be applicable to the Mediterranean landscape if another spatial scale of resolution is considered. After all, the homogeneous character of the 'matrix' is always defined relatively and is highly dependent on the scale of data registration. If comparative analyses of landscape pattern dynamics is to succeed, the problems related to the establishment of a common set of concepts and a common methodology will have to be considered further.

The definition of discrete entities (e.g. landscape elements) for analysis also deserves attention. Definitions vary according to academic discipline, as in the case of hedgerow networks. For example, an ecologist will consider a hedge-row to be an 'homogeneous' ecological unit, as defined by tree cover or tree species. An agronomist will, regardless of the structure, define a hedgerow as a field boundary: it is considered to be a linear element bordering a given field; it is a management unit; its structure is of no interest. In interdisciplinary studies, landscape elements, as mappable elements, need to be defined similarly for all disciplines. If the objective is to study the effects of management, management units will be preferred (Baudry *et al.*, 1993).

Delimitation and comparison of landscapes

The size and delimitation of the study object – the landscape – also needs attention. Many well-established, theoretical considerations which concern the delimitation and definition of geographical regions will be relevant in the present context.

In general, the region is defined as a part of the Earth's surface, which is homogeneous with respect to some (or one) criteria which define the limits of the region. A generally accepted typology of regions differentiates between 'functional' and 'formal' regions. The 'functional' region is delimited and defined on the basis of the functional coherence within the region (examples would be: a farm and its fields; a catchment area of a stream; a city and its hinterland). The 'formal' region is homogeneous with respect to one or more characteristics (variables). Thus, the division into formal regions is equivalent to a classification procedure. Meaningful regions and analytical results are totally dependent on the selection of discriminating values for the selected variables.

Scientists from various relevant disciplines agree upon the utility of regionally-based studies. Gulinck (1986) notes that 'landscape ecology is concerned with land units and suggests a kind of – *spatial* (our addition) – limitation'; Haggett (1983) states that 'ecological analysis in geography concerns itself with the study of connections between human and environmental variables. In this type of analysis we are studying the relation within particularly bounded geographical spaces (e.g. *regions*, our addition)'. This, however, still leaves a very important problem: how do we define and delimit a study object in an acceptable way that ensures its uniqueness? This problem has not only to do with the size but also the delimitation criteria and the type of region. Here it is important to remember that the region exists only as an intellectual concept that is useful for a particular purpose (see e.g. Holt-Jensen, 1988).

The analysis of statistical information often leads to research within the field of landscape analysis, agriculture and the environment becoming based on 'administrative' regions. In most cases these regions are neither 'functional' nor 'formal' regions in the senses defined above; but they are relevant and may even be inevitable tools in the investigations. Indeed 'administrative' regions might offer the only realistic basis for statistical information on an historical scale, including up-to-date variables.

In landscape pattern analysis we have to cope not only with the problem of delimiting relevant research objects (regions) but often we have to face the fact that the interdisciplinary data requested are not available for comparable units (Rasmussen *et al.*, 1995). Figure 2.10 provides an example from an analysis which aimed at relating land-use patterns to nutrient circulation within a watershed.

CONCLUSION

There are numerous examples of regional studies on landscape change and land-use transformation, but comparative studies are scarce. Approaches to the diversity of European landscapes do not often deal with the mechanisms of differentiation. In this chapter we have identified those crucial issues which should be considered in the process of defining a suitable analytical framework. Key concepts in progressing towards the building of comparative models of change have also been briefly touched upon.

It must be concluded that any framework for comparative landscape pattern analysis must encompass a variety of spatial scales – from the field level (where techniques are applied to maintain a certain land use), to the household level (where decisions are taken), to the regional and country level (where policies are implemented). The focus on scale and on spatial aspects of phenomena and processes will be central to a successful analytical framework.

Time-series analysis is a natural part of studies which deal with landscape pattern dynamics. In this context, too, the selection of scale deserves considerable attention because it will influence analytical results. Important fluctuations might be overlooked, and the estimated order of magnitude for the changes might vary significantly as a result of the definition of the analytical framework.

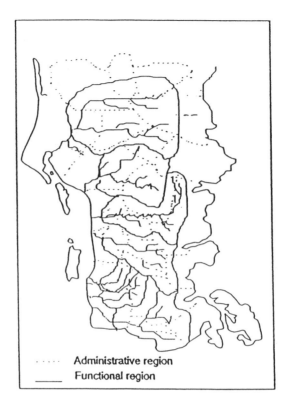

Figure 2.10 Administrative boundaries and watershed boundaries in a part of Denmark. Lack of correspondence between boundaries will cause data handling problems in many interdisciplinary landscape studies that wish to combine socioeconomic data with parameters describing environmental conditions (from Rasmussen *et al.*, 1995)

Nevertheless, we do not intend to provide a rigid set of methods or solutions. Several valuable research contributions have been developed and made available from various scientific disciplines; yet, there is still a need to assess them and test their efficiency and robustness within an interdisciplinary framework. Even so, the diversity of theories, as well as topical focus, in the various disciplines that deal with rural landscape changes remains problematic. Theories need to be tested against data and alternative explanations for a given set of data need to be investigated. Theories mirror the particular beliefs and values of scientists and our approach is not necessarily compatible with those which have no spatial content or which do not include a combination of different spatial scales of analysis. Therefore, there is an urgent need for clarification between different research groups on how they can make progress on this main task. Practical case studies might be the most suitable point of departure.

REFERENCES

Allen, T. H. F. and Starr, T. B. (1982). *Hierarchy: perspective for ecological complexity.* The University of Chicago Press, Chicago and London, pp. 310

Andriesse, W., Fresco, L. O., van Duivenbooden, N. and Windmeijer, P. N. (1994) Multi-scale characterization of inland valley agroecosystems in West Africa. *Netherlands Journal of Agricultural Science,* **42**: 159–79

Barrensoe, B. (1987). *En foreloebig kortlaegning af tilplantede landbrugsarealer I dette aarhundrede.* Marginaljorder og Miljoeinteresser, Miljoeministeriets Projekt-undersoegelser 1986, Teknikerrapport no. 3, Copenhagen

Baudry, J. (1992). Introduction Générale. In Auger, P., Baudry, J. and Fournier, F. (eds) *Hiérarchies et échelles en écologie.* Naturalia Publications, Paris pp. 9–18

Baudry, J. (1992). Dépendence d'échelle d'espace et de temps dans la perception des changements d'utilisation des terres. In Auger, P., Baudry, J. and Fournier, F. (eds) *Hiérarchies et échelles en écologie.* Comité Francais SCOPE, Naturalia Publications, pp. 101–14

Baudry, J. and Burel, F. (1990). Structural dynamic of a hedgerow network landscape in Brittany France. *Landscape Ecology,* **4**: 197–210

Baudry, J. and Bunce, R. G. H. (eds) (1991). *Land abandonment and its role in conservation.* Options Méditerranéennes, Zaragoza, A15, pp. 148

Baudry, J., Tatoni, T., Luginbühl, Y., Barre, V. and Berlan-Darqué, M. (1993). Bocages et Environnement. *Recherches, Etudes, Environnement, Développement,* **41–42**: 15–19

Baudry, J. and Tatoni, T. (1993). Changes in landscape patterns and vegetation dynamics in Provence, France. *Landscape and Urban Planning,* **24**: 153–9

Baudry, J., Thenail, C., Le Coeur, D., Burel, F. and Alard, D. (1993). *Landscape ecology and grassland conservation.* Grassland management and nature conservation. Occasional Symposium No. 28. British Grassland Society, Aberystwyth, pp. 157–66

Bazin, G. and Larrère, G. R. (1983). Du système agropastoral à la spécialisation laitière. In *Système agraire et pratiques paysannes dans les Monts Dômes.* INRA Publications, Paris pp. 1–156

Berglund, B.E. (1991). The cultural landscape during 6000 years in southern Sweden – The Ystad Project. *Ecological Bulletin,* **41**: 1–495

Breuning-Madsen, H., Reenberg, A. and Holst, K. (1990). Mapping potentially marginal land in Denmark. *Soil Use and Management,* **6**(3): 114–20

Burel, F. (ed.) (1995). Ecological patterns and processes in European agricultural landscapes. Special issue of *Landscape and Urban Planning,* **31**(1–3): pp. 412

CEC (Commission of the European Communities) (1992). *The European Commission's guidelines for the reform of the CAP.* Europe Documents, Commission of the European Communities, Bruxelles, 1689: 1–8.

Dumanski, J., Pettapiece, W. W., Acton, D. F. and Claude, P. P. (1993). Application of agro-ecological concepts and hierarchy theory in the design of databases for spatial and temporal characterisation of land and soil. *Geoderma,* **60**: 343–58

Forman, R. T. T. (1995). *Land Mosaic: The ecology of landscapes and regions.* Cambridge University Press, Cambridge, pp. 632

Forman, R. T. T. and Godron, M. (1986). *Landscape Ecology.* John Wiley & Sons, New York, Cambridge pp. 619

Fresco, L. O. and Kroonenberg, S. B. (1992). Time and spatial scales in ecological sustainability. *Land Use Policy,* **9**: 155–68

Fresco, L. O., Stroosnijder, L., Bouma, J. and van Keulen, H. (eds) (1994). *The future of the land: Mobilising and integrating knowledge for land use options.* John Wiley & Sons, Chichester, pp. 409

Gilibert, J. (1993). De la jachère et des autres solutions. *Le Courrier de l'environnement de l'INTRA*, no. 19

Gulinck, H. (1986). Landscape ecological aspects of agroecosystems. *Agriculture, Ecosystems and Environment*, **16**: 79–86

Haber, W. (1990). Using landscape ecology in planning and management. In Zonneveld, I. S. and Forman, R. T. T. *op. cit.,* pp. 217–32

Haggett, P. (1983). *Geography, a modern synthesis.* Harper and Row, NY, pp. 218

Haycock, N. E. and Muscutt, A. D. (1995). Landscape management strategies for the control of diffuse pollution. *Landscape and Urban Planning*, **31**: 313–21

Holt-Jensen, A. (1988). *Geography – History and Concepts.* Paul Chapman Ltd., London, pp. 186

Laurent, C. (1991). Place de l'activité agricole dans l'espace, l'exemple d'une région agricole de Basse-Normandie, le Pays d'Auge. *Economie Rurale*, **202–203**: 34–9

O'Neill, R.V. (1989). Perspective in hierarchy and scale. In Roughgarden, J., May, R. M. and Levin, S. A. (eds) *Perspectives in ecological theory.* Princeton University Press, Princeton, pp. 140–56

Pinto-Correia, T. (1993). Landscape Monitoring and Management in European Rural Areas: Danish and Portuguese Case Studies of Landscape Patterns and Dynamics. PhD Thesis. *Geographica Hafniensia* A1, Copenhagen, pp. 165

Rabbinge, R., van Diepen, C. A., Dijsselbloem, J., de Koning, G. H. J., van Latesteijn, H. C., Woltjer, E. and van Zijl, J. (1994). Ground for choices: A scenario study on perspectives for rural areas in the European Community. In Fresco, L. O., Stroosnijder, L., Bouma, J. and van Keulen, H. (eds) *The future of the land: Mobilising and integrating knowledge for land use options.* John Wiley & Sons, Chichester, pp. 93–121

Rasmussen, M., Reenberg, A. and Bartholdy, J. (1995). Nitrogen fluxes from Landscapes – Comparison on watershed level. In Ryszkowski, L. and Balazy, S. (eds) *Functional Appraisal of Agricultural Landscapes in Europe.* EUROMAB and INTERCOL, Polish Academy of Science, Poznan, pp. 19–30

Rayner, S. (rapporteur), Bretherton, F., Buol, S., Fosberg, M., Grossman, W., Houghton, R., Lal, R., Lee, J., Lonergan, S., Olsen, J., Rockwell, R., Sage, C., van Imhoff, E. (1994). A wiring diagram for study of land use/land cover change. In Meyer, W. B. and Turner, B. L. (eds) *Changes in land use and land cover: a global perspective.* Cambridge University Press, Cambridge, pp. 13–53

Reenberg, A. (1990). Rural Land use pattern in Denmark. Recent regional changes as an indicator for future development. In Brossier, J. (ed.) *AGRICULTURE – Methods and socioeconomic criteria for the analysis and the provision of land use and land evaluation.* EEC Report EUR 12340 en, Luxembourg, pp. 169–82

Reenberg, A. (1992). Guidelines for a comparative research on land use changes and their environmental impact in rural areas in Europe. In Proceedings from MAB workshop in Kiev 3–7 June 1991, Ukrainian MAB Committee: Baudry, J., Bure, F. and Hawrylenco, V. (eds) (1992). *Comparison of Landscape Pattern Dynamics in European Rural Areas.* 1991 Seminars. MAB/UNESCO, France/Ukraine, pp. 8–20

Reenberg, A. (1993). Changing rural landscapes: A holistic approach to processes driving environmental changes. Invited contribution to Council of Europe, *The state of the environment in Europe: the scientists take stock of the situation.* Council of Europe, Milan 1993, pp. 309–11

Reenberg, A. (1996). A hierarchical approach to land use and sustainable agriculture in the Sahel. *Quarterly Journal of International Agriculture*, **35**(1): 63–77

Reenberg, A. (1998). Analytical approaches to agricultural land use systems in the Sahel. *SEREIN Occasional Papers No. 8*. Copenhagen, pp. 171

Reenberg, A. and Fog, B. (1996). The spatial pattern and dynamics of a Sahelian agroecosystem – Land use systems analysis combining household survey with georelated information. *GeoJournal* **37**(4): 487–99

Reenberg, A. and Pinto-Correia, T. (1994). Rural Landscape Marginalization. Can General Concepts, Models and Analytical Scales be Applied throughout Europe? In *Analysis of Landscape Dynamics – Driving Factors Related to Different Scales*. Proceedings MAB/UNESCO. UFZ Leipzig-Halle, pp. 3–18

Skov-og Naturstyrelsen (1987). *Marginaljorder og miljoeinteresser – en sammen-fating*. Miljoeministeriet, pp. 276

Stromph, T. J., Fresco, L. O. and van Keulen, H. (1994). Land Use System Evaluation: Concepts and Methodology. *Agricultural Systems*, **44**: 243–55

Stryg, P. E. (1992). *EF-Landbrugsreformens konsekvenser for udviklingen I dansk landbrug frem til 2005*. Marginaljordene og landskabet, Proceedings of the Danish IALE seminar, Copenhagen, 25 September, pp. 1–8

Thenail, C. B. J. (1996). *Consequences on landscape pattern of within farm mechanisms of land-use changes (example in western France)*. Land use changes in Europe and its ecological consequences, European Centre for Nature Conservation, Tilburg, pp. 242–58

Turner II, B. L. and Meyer, W. B. (1994). Global land use and land cover change: an overview. In Meyer, W. B. and Turner, B. L. (eds) *Changes in land use and land cover: a global perspective*. Cambridge University Press, Cambridge pp. 3–10

van Diuvenbooden, N. (1995). *Land use systems analysis as a tool in land use planning*. Wageningen Agricultural University, pp. 176

Vos, W. and Stortelder, A. (1992). *Vanishing Tuscan Landscapes – Landscape ecology of a Submediterranean-Montane area (Solano basin, Tuscany, Italy)*. Pudoc Scientific Publishers, Wageningen, pp. 404

Wiens, J. A. (1989). Spatial scaling in ecology. *Functional Ecology*, **3**: 385–97

Young, M. D. and Solbrig, O. T. (1993). *The World's Savannas. Economic driving forces, ecological constraints and policy options for sustainable land use*. Man and Biosphere vol. 12. Parthenon and UNESCO, Casterton and Paris, pp. 350

Zonnenveld, I. S. and Forman, R. T. T. (eds) (1990). *Changing Landscapes: an ecological perspective*. Springer-Verlag, New York, pp. 286

CHAPTER 3

DESCRIPTION AND ANALYSIS OF THE NATURAL RESOURCE BASE

O. Bastian

INTRODUCTION

Although landscapes differ throughout Europe, similar and unifying concepts for the description of landscapes and their development are needed. Methodological approaches need to be similar, irrespective of the peculiarities of special regions or countries. The present chapter will look into a number of important issues with respect to *landscape ecological analysis* – analysis, diagnosis and prognosis, definition of natural potentials/landscape functions, the indicator principle, landscape evaluation, the work in different scales or dimensions, the definition of landscape stability and further specific questions of research and working steps. An approach for the investigation of landscape change from a landscape ecological point of view is described and recommended.

The application of a standardized methodology for landscape ecological analysis allows a comparison of common features, peculiarities and trends of regions. This will improve the scientific basis for investigations of landscape change, which today still lack a homogeneous methodology applicable in all test areas. This is particularly true for the definition of scientific objectives, the choice of analytical data (indicators), the processing and interpretation of information, and the communication of results.

THE LANDSCAPE ECOLOGICAL APPROACH

The framework of landscape research and planning

At present, in many countries, there are two basic approaches to landscape analysis (Bastian and Sandner, 1991). Landscape architects prefer an artistic–aesthetic approach and favour the relatively small-sized designing of parks, green zones and other landscape details. On the other hand, the scientific approach, mainly represented by geographers and ecologists, focuses on landscape, society and the interactions between landscape and society. The latter approach will be put at the centre of the following discussion.

The starting point to the discussion is provided by viewing any landscape as 'a part of the Earth's surface characterized by a relatively homogeneous structure and similar processes' (Neef, 1967) or as a result of the coincidence of natural conditions (with the components relief, soil, water, climate, vegetation) and human influences (land use, management techniques).

43

In Germany, the Saxonian school of landscape ecology has arrived at highly relevant theoretical and practical results in their approach to modern landscape research and planning (e.g. Neef, 1967; Haase, 1978, 1985, 1990; Mannsfeld, 1979; Niemann, 1982, Haase *et al.*, 1991). Following Haase (1990), the following approach has been developed:

(1) Methodological bases for the structural, functional and conflict analysis of landscapes, as well as the multiple-step analysis of the economic and non-economic evaluation of interactions between society and nature.
(2) Derivation and interpretation of 'natural potential' as a basis for an assessment of the resources in a region.
(3) Determination of the stability, resilience, and carrying capacity of landscapes.
(4) Methods of transferring the results of the inventory and survey of landscapes into the planning process.
(5) Methods for the polyfunctional assessment of performance, suitability, and resilience of landscape elements and units by an optimality approach.

During the past few years, this landscape research approach has resulted in the development of a coherent, highly consistent concept of landscape ecological research and planning.

Figure 3.1 shows the general approach to landscape ecological research and planning which essentially involves the analysis of landscape change. Beginning on the left hand side of the figure, this approach deals with landscape development over time – from an historical state, via a present state, to a future state which cannot be foreseen as yet in detail. In view of rapid landscape changes, this consideration of the time factor is very important. Starting with knowledge of historical structures and processes, as well as knowledge of recent trends, future landscape states or environmental problems can be predicted. Thus the first aim of *landscape analysis* is the reconnaissance of the natural landscape and land use; that is to say, the characterization of the recent ecological state of the landscape.

To further this aim, the use of indicator variables (landscape characteristics) is necessary and the indicator principle has become the basis of several methods of environmental observation (i.e. *landscape ecological and biomonitoring*). Areas with corresponding or similar natural indicators and human influence can be combined to identify landscape units (*landscape synthesis*). *Landscape diagnosis* deals with the comparison of landscape characteristics with judgements as to value, which are formulated by human society, and so-called targets/guidelines or conceptual ideas of landscape development. From this, the characterization of landscape functions (suitability, performance, resilience) can be derived, as well as assessments made of landscape change, threats to natural resources and resource-use conflicts.

These results provide a scientific foundation for *landscape management* (i.e. to all measures in human society aimed at maintaining and improving

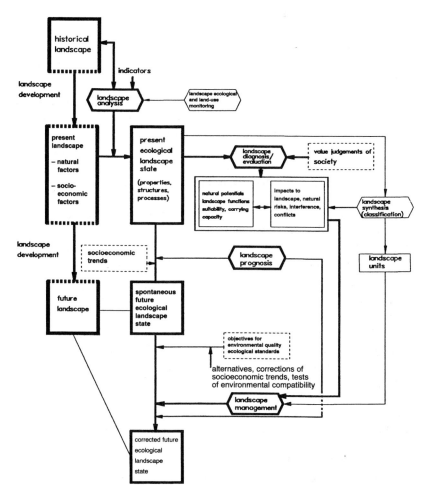

Figure 3.1 Concept of landscape ecological research and planning with special consideration of landscape change (from Bastian, 1991; design: Kiessling)

landscape structure and function). If socioeconomic processes proceed irregularly and environmental consequences are not taken into consideration enough in economic projects, in the future an ecologically unsatisfactory 'spontaneous ecological state' will appear. However, in many cases environmental objectives, standards, etc. exist; also, proposals which have far reaching impacts on the landscape usually have tests of environmental compatibility before they are carried out. Consequently, adaptations to plans and the choice of alternatives are often necessary and possible. In this way, landscape development can be directed towards an 'optimized future ecological state'. The aim of several methods of *landscape prognosis* is to compare different scenarios of future landscape development and to predict the efficiency of landscape management.

Natural potentials/landscape functions

The concept of 'natural potential', which was first used by Neef (1966) in a geographical context, and further developed by Haase (1978), Mannsfeld (1979) and others, is a useful approach to the analysis and assessment of landscape and its changes for human utilization. The terms 'natural potential'/'landscape function' (both are often applied synonymously) describe ecological functions ('balance of nature') as well as the ability of the landscape to satisfy the needs and demands of human society in a broad sense. Landscape function can be classified into economic, ecological and social functions (see Table 3.1).

The following natural potentials/landscape functions should be considered when characterizing 'balance of nature', its changeability and need for protection:

(1) biotic regulation and regeneration potential (especially habitat value with regard to species and biotopes);
(2) groundwater recharge, the functions of surface waters, runoff regulation;
(3) yield and decontamination potentials, resistance of soils to water and wind erosion (the productive and regenerative functions of soils);
(4) air hygienic and microclimate balancing functions;
(5) recreation potential, including the aesthetical and ethical values of landscapes.

Figure 3.2 shows the variation of natural potentials/landscape functions by way of an example of groundwater recharge in a Saxon test region north of Dresden. The groundwater recharge depends on geological, pedological and climatic parameters (precipitation, evaporation), on relief (slope) and land use. From the 1930s (Figure 3.2a) to the 1980s (Figure 3.2b), land-use changes (surface sealing and drainage) have caused a reduction of groundwater recharge in several areas of the test region.

The indicator principle

For the analysis of landscapes, it is very useful to select appropriate indicators (landscape characteristics) out of the multitude of landscape attributes. These indicators should:

(1) be easily comprehensible and well-defined;
(2) possess significance as important landscape ecological features;
(3) characterize dynamic aspects (e.g. changes of vegetation, land use, water balance, ground water level).

Table 3.1 Classification of landscape functions (suitability, performance, resilience), arranged in a hierarchical order

Groups of functions

A   production (economic) functions
- availability of renewable resources
  production of biomass (suitability for cultivation)
    plant biomass
      arable fields (husbandry)
      permanent grassland
      special crops (e.g. fruit culture)
      wood (forestry)
      animal biomass
      game (hunting)
      fish (fishing, pisciculture)
    water accumulation
      surface waters
      groundwater
- availability of non-renewable resources
      mineral raw materials, building materials
      fossil fuels

B   ecological functions
- regulation of matter and energy circles
    pedological functions (soil)
      resistance against erosion
      resistance against underground wetness
      resistance against drying up
      resistance against compaction
      decomposition of harmful matters (filtering, buffering and transforming functions)
    hydrological functions (water)
      groundwater recharge
      water storage/run-off balance
      self-purification of surface waters
    meteorological functions (climate/air)
      temperature balance
      enhancing of atmospheric humidity
      influencing of wind
- regulation and regeneration of populations and communities (of plants and animals)
    biotic reproduction and regeneration (self-renewal and maintenance) of biocoenoses
    regulation of organism populations (e.g. pests)
    conservation of the gene funds

C   social functions
- psychological functions
    aesthetic functions (scenery)
    ethical functions (gene funds, cultural heritage)

Continued

Table 3.1 *Continued*

- information functions
    - functions for science and education
    - (bio-)indication of environmental situation
- human–ecological functions
    - bioclimatical (meteorological) effects
    - filtering and buffering functions (chemical effects – soil/water/air)
    - acoustic effects (noise control)
- recreative functions (as a complex of psychological and human–ecological effects)

It is possible to classify indicators in varying ways: for example, by landscape function, by spatial scale of their significance, and by measured attribute. Also, interval, ordinal or nominal information can be used. Figure 3.3 and Table 3.2 schematically express the use of different indicators for the characterization of natural potentials/landscape functions.

Ecological evaluation

The most important step in analysing and interpreting landscapes, especially with regard to dynamic aspects, is their 'ecological evaluation'. Here we leave the level of observation or statistical measurement of the landscape, and its changes, and create the basis for landscape management. Generally, an evaluation is a relation between a valuing subject and an object being valued in relation to some given criteria (Bechmann, 1989). Thus an 'ecological evaluation' is a valuation of spatial structures, utilizations, functions and potentials with regard to the functioning of the 'balance of nature'. Valuation procedures arrange and regulate valuation events concerning form and content.

A *universal algorithm* for landscape ecological evaluations includes the following steps:

(1) definition of the aim of evaluation;
(2) choice of suitable methods;
(3) definition of the criteria for the evaluation and the scale of analysis;
(4) analysis of necessary data;
(5) weighting and combination of the analysed data;
(6) interpretation of the results with regard to measures.

The problem of scale or dimension

Overcoming the scale (dimension) problem – i.e. the choice of an appropriate spatial scale of analysis, which includes relevant landscape objects (as indicators) and methods of measurement – is a fundamental prerequisite for the success of geographical and landscape ecological research, as well as its practical use, mainly in landscape planning. It is always necessary to consider

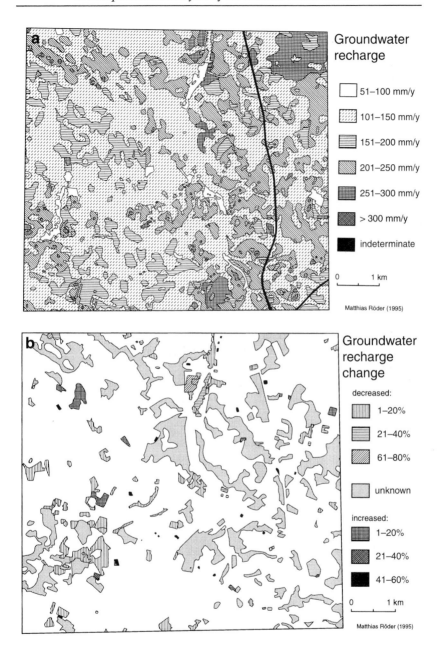

Figure 3.2 Change of groundwater recharge in a test area around Moritzburg north of Dresden between the 1930s (Figure 3.2a) and the 1980s (Figure 3.2b – only test lots with a change of groundwater recharge are marked; on numerous lots the extent of the change is unknown) (investigation and design Röder, 1995, unpublished)

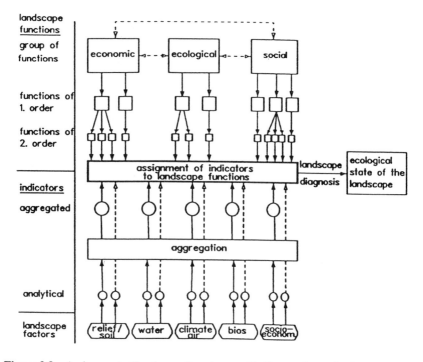

Figure 3.3 Assignment of landscape functions and indicators for evaluating the ecological situation of landscapes (from Bastian, 1991, 1992; design: Kiessling)

the object size (landscape elements), the anticipated content, the detailed method used for analysis and assessment, and the scale of (cartographical) representation. For example, the transition from a large to a small scale causes a loss of detail, but allows an increasing concentration on the essential structures, processes and relationships in the landscape.

Examples of the facts and objects to be handled in different dimensions and scales are: the ecological spatial units, the system of landscape plans, the hierarchical order of vegetation units, biotopes and their linking ecotones, the degrees of naturalness of vegetation, the types of stability and diversity (see Figure 3.4).

Taking into account the availability of data, the available time and staff expense and the desired quality of the results, an approach within a hierarchical system of several indicators and methods can be chosen.

THE ASPECT OF TIME: LANDSCAPE CHANGES

Common trends of landscape change in Europe

The investigation of landscape change has, above all, two objectives (Bastian, 1987):

Table 3.2 Landscape characteristics (indicators), which are necessary for the assessment of landscape functions (natural potentials) in landscape diagnoses

Indicators	y (s)		y (e)		p		g		r		b		re	
Scale (dimension)	m	l	m	l	m	l	m	l	m	l	m	l	m	l
geological basis					(x)		(x)		(x)		(x)	(x)		
relief														
slope	x	(x)	x	(x)			x		x				x	(x)
differences in altitude													x	
small structures													x	x
soil														
substrate peculiarities	x	x	x	x	x	x	x	x	x	x	(x)	(x)		
main soil forms	x				x		x		x					
soil forms (pedotopes)		x		x		x		x		x				
water														
existence/quality of surface waters									(x)	(x)	x	x	x	x
depth of ground watertable/water balance of the soils	x	(x)			x	x	x		x	x	(x)	(x)		
climate														
annual precipitation	x						x							
monthly preciptiation			x				(x)							
potential evaporation							x							
annual temperatures	x													
occurrence of frost	x													
biota														
biotope types											(x)	x		x
vegetation units		(x)										x		(x)
habitat structures												x		
species									x			x		(x)
spatial parameters											(x)	x		
potential natural vegetation	(x)	(x)									x	x		
land use														
land-use types	x		x	x	(x)	(x)	x	x	x	x	x	x	x	x
landscape elements			(x)	x					x	x		x	(x)	x
specific information (surface sealing, irrigation/drainage, rotation of crops, use of fertilizers)	x				x	x			(x)		(x)			

y - biotic yield potential (s - suitability, e - sensitivity: water erosion); p - groundwater protection; g - groundwater recharge; r - regulation of surface runoff; b - biotic regulation potential (habitat function); re - recreational potential; m - medium and small scales (about 1:25,000 and smaller); l - large scales (about 1:10,000 and larger)

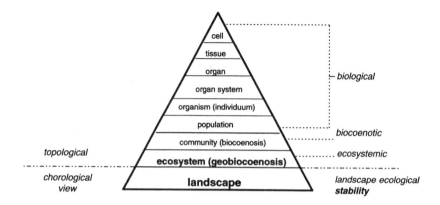

Figure 3.4   Hierarchical sequence of biological/ecological systems, and the differentiation of stability (from Bastian, 1991, 1992, modified)

(1) The early recognition and assessment of landscape changes makes corrective intervention possible; through timely regulation, undesired processes of change can be counteracted at relatively low economic expense.
(2) The knowledge and documentation of the ecological situation of bygone epochs are part of the preservation of our historical and cultural heritage.

In view of an increasing speed and intensity, the investigation of landscape changes has become more and more important; this is particularly the case when landscape change threatens the sustainable development of human society.

A number of fundamental trends in landscape changes in Central Europe can be identified which have been caused by man since his early settlement of the continent (Bastian and Bernhardt, 1993). The landscape changes have been brought about almost exclusively by material and technological advances, and thus by social developments. The periods of time occupied by each of a number of stages of development have successively become shorter, in an almost logarithmic sequence. Indeed, this acceleration in the pace of man-made influences and impacts has made it more and more difficult for natural processes to become stabilized. As a result, the landscape balance as a whole has been subject to destabilization at an increasing rate. At first, environmental degradations were only local and limited; then, spread to larger regions, and by now have reached global dimensions. There has been a rapid increase in the proportion of landscapes which have suffered from irreversible change.

More recently, because of the marginalization of agriculture, the almost free availability of energy and nutrients, and the complexity of structures and processes, together with their global interconnection, the landscape has increasingly lost its significance as a reflection of the human society of a particular region (Muhar, 1995). In addition, although landscape changes follow common regularities in different regions, there are many specific peculiarities: different landscapes with different resources react in a variety of ways to the same type of human activity.

Current landscape development is characterized by the intensive use of almost all resources. Nowadays all landscapes are exposed to anthropogenic material and energy throughputs at levels many times higher than in the past, with substances extraneous to nature becoming omnipresent and pollution spreading to large areas. There is also a rapid increase in the multiple use of landscapes which leads to conflict and neighbourhood effects. In general, biodiversity is strongly reduced.

The future development of European agricultural landscapes thus depends on complex economic and political conditions and, to a great extent, is unforeseeable.

Ecological stability and instability

The analysis and evaluation of a landscape, as described above, begins with a static view. At first landscape functions at a fixed time (t_1) are assessed; then, in order to register landscape changes, the landscape must be investigated at least a second time (t_2). Thus the landscape change is expressed by the difference between these two stages $(t_1 - t_2)$.

In order to decide if a landscape change has taken place within a certain time, the concept of *ecological stability* is helpful. According to Gigon (1984), ecological stability characterizes the maintenance of an ecological system and its ability to return to the starting point after changes. Concerning human influences, a landscape is stable if no irreversible disturbances of natural potential take place as a result of human impact (Buček *et al.*, 1985). In practice, however, the imprecise, partly diverging and thus confusing use of the term 'stability' has proved problematic. There is no stability per se, rather stability exists only with regard to a certain purpose, in which different criteria and limits of tolerance, partly determined by human society, play an essential part (Figure 3.4).

Closely connected with the concept of ecological stability is the question of resilience – the capability of a landscape to react to human impacts (or natural influences) in such a manner that the preservation of its state is possible (Neumeister, 1984). According to this concept, 'ecological carrying capacity' marks the limit for a possible resource use – the level of loading (utilization) that can be tolerated. Once a particular carrying capacity is exceeded, *ecological instability* marks the non-maintenance of an ecological system with the inability to return to the starting point after changes.

Specific questions in research on landscape changes

With regard to the investigation of landscape changes, the *significance* of changes must be investigated. There are three possibilities:

(1) No changes are detectable.
(2) Changes are detectable but are insignificant from both landscape ecological and economic points of view (e.g. low-level soil erosion).
(3) Significant changes are detectable in landscape structures and functions (e.g. ploughing of meadows rich in plant species; control of running water).

Other problems are:

(1) The *speed* of changes from t_1 to t_2. What is the spatial and temporal *distance between the causes and the effects* of changes?
(2) *The spatial dimension of changes.* At which level of ecosystems or landscapes are the changes taking place and are they significant? Are the changes locally bound, singular cases or widespread common phenomena (the *singularity* of changes)?
(3) The *reversibility or irreversibility*. Are the changes reversible? If yes: at what expense and over what time period?
(4) The *acceptance* of changes *by human society*. Are landscape planning guidelines, environmental quality standards and landscapes changes compatible with sustainable development?

Working steps for the investigation of landscape changes

An investigation of landscape changes should proceed as outlined below. First, *characteristic forms of change* which will be considered more comprehensively, *suitable methods* for analyses/diagnoses/evaluations, necessary *indicators* and representative *test areas* (with a good data availability) should be agreed.

Second, data for the *initial* and *contemporary landscapes* should be collected (i.e. t_1 and t_2).

Third, the interpretation of the current *social and economic changes* with regard to landscape ecological conditions as well as the analysis of the landscape changes and their interpretation concerning *landscape functions* should be carried out.

Fourth, the *driving forces* and effective mechanisms/causal connections should be identified (i.e. causal relationships between land-use changes and landscape ecological changes), as well as an assessment of landscape changes with regard to their ecological carrying capacity and their *acceptance* by human society (the problem of guidelines).

Finally, information on *trends*/prognoses/manageability for future landscape development should be derived.

'BIOTIC REGULATION POTENTIAL' AS AN EXAMPLE

Biotic regulation potential characterizes the capability of landscape units and elements to maintain living processes, biodiversity and complexity, as well as the stability of ecosystems (Haase, 1978). These concepts will now be described in more detail.

The assessment of biotic regulation (and regeneration) potential is still a rather difficult methodological problem. Although we know about the complexity of structures and processes within nature, rather simple but sufficiently reliable evaluation methods, procedures for landscape planning, ecosystem management, and environmental impact assessment need to be made available. These methods must be adaptable to special purposes – i.e. their objectives, dimensions, precision, available data, time and labour. But the lack of landscape ecological information is proving to be a bottleneck in ecological planning projects; this is particularly true for biotic landscape elements such as flora, fauna and biotopes because they change permanently and very quickly. A possible approach to solve these problems is the analysis and diagnosis of the landscape with the help of a network of hierarchical test areas and methods varying in their scale and intensity of examination (Bastian, 1990, 1991, 1992; Bastian and Haase, 1992, Tables 3.3, 3.4, Figure 3.5). Similar or analogous methodological frameworks can be elaborated for the investigation of other natural potentials/landscape functions.

In order to determine meaningful indicators, significant landscape features – concerning biotic regulation capacity – must be selected consisting of certain plant and animal species, vegetation types, habitat structures and biotopes. In essence, the analysis and interpretation of habitat values can be obtained through studies at five (sometimes six) levels of investigation. If necessary, these levels can be further subdivided.

Levels 1 and 2 do not require fieldwork but can be investigated through surveys on large territories (e.g. countries, counties, regions), or smaller areas (e.g. rural districts, landscape conservation areas) using existing data (satellite and aerial pictures, topographical and thematical maps).

For levels 3, 4 and 5, as a rule, it is necessary to analyse biotic landscape elements directly in the field, which is possible only in rather small test areas. Exactness and expense increase from land use and biotope type mapping (degree 3), via vegetation mapping (mainly in the range of associations, sub-associations and vegetation forms) of flora and fauna (degree 4), to the analysis of organism groups which are rather difficult to record (e.g. soil fauna, arthropods, fungi – degree 5).

Very detailed investigations of biological objects (e.g. morphometrical measurements and biochemical/ecophysiological laboratory analyses) correspond to a sixth level. However, these investigations belong to the more non-spatial discipline of bioecology and, according to Leser (1989), give the necessary specialist knowledge for geographical approaches but are not at the centre of the landscape ecological approach.

Table 3.3 Scale ranges, test areas and approaches for evaluating landscape habitat value/the biotic regulation potential (from Bastian, 1991, 1992)

Level of research	Test area	Scale	Approaches
no fieldwork			
1 a	country	1:200,000	interpretation of natural factors (environmental media, pollution), land use (pattern, conflicts)
1 b	district	1:25,000	analysis of biotope-linking, assessment of floristic and faunistic maps
2	parts of a district, communities	1:10,000	as level 1, but a more detailed resolution level
fieldwork			
3	as in 2	1:10,000 and larger	biotope mapping (biotope types, landscape elements), land-use analysis (detailed)
4	small test areas	as in 3	analysis of actual vegetation (plant communities, vegetation forms, indicator species) and the location of landscape elements/ biotopes or habitats
5	as in 4	as in 3	registration of groups of species being difficult to detect (irregularly appearing, mobile, living in concealment) or hardly determinable groups of species (especially animals)
mainly laboratory analysis 6	pointwise sampling	as in 3	morphometrical and biochemical (ecophysiological) investigations (especially within biomonitoring programmes)

Analytical indicators can be interpreted using the following criteria:

(1) degree of naturalness of vegetation;
(2) regenerative capacity, age, duration of development;
(3) diversity;
(4) spatial criteria (biotope size, isolation, connectivity);
(5) representativeness;
(6) rarity, degree of threat.

A number of these criteria suffer serious scientific/methodological problems and are not undisputed. Looking at them in more detail:

Table 3.4 Selected methodical approaches for evaluating the landscape's biotic regulation potential (especially landscape habitat value) in different dimensional stages (see Table 3.3) (from Bastian, 1991, 1992)

Criteria/ Indicators	Scale/Level of investigation				
	1	*2*	*3*	*4*	*5*
ecological stability	interpretation: land use and its interferences, situation of environmental media (esp. air and water pollution)		interpretation of hemeroby*/naturalness of vegetation		
rarity/degree of threat	analysis of small-scale floristical and faunistical maps		singularity/ replacability of biotope types	rare/en- dangered plant species and communities	rare/en- dangered (plant)/ animal species
			suitability of biotope types for rare species	suitability of single biotopes for rare species	
degree of naturalness	dominance and mosaic types of naturalness of areas	coefficient of landscape balance; index of ecological value	degree of naturalness of vegetation/ hemeroby	naturalness of plant communities	
regenerative capacity (age, duration of development)	interpretation of land use		age, duration of development of biotope types		
diversity	interpretation: land-use, impacts of environmental media				
species diversity			rank of habitat types (concerning species richness)	number of plant species	number of (plant) /animal species
spatial diversity		land-use diversity	landscape elements, diversity of biotope pattern	diversity of mosaic of vegetation units (associations)	

Continued

Table 3.4 *Continued*

Criteria/		Scale/Level of investigation			
Indicators	*1*	*2*	*3*	*4*	*5*
spatial aspects (biogeography)	regional biotope linking	minimum areas; edges, critical distances between: land-use types/biotope types/biotopes local biotope linking		vegetation units	
complex biotope values		biotope value of land-use types	biotope value of biotope types	biotope value of single biotopes	

* hemeroby: extent of human influence on vegetation cover

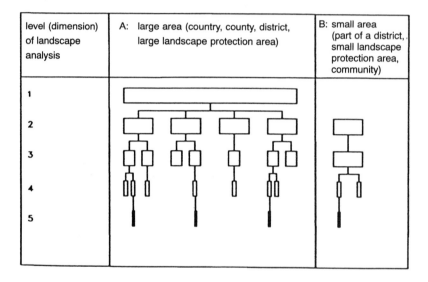

Figure 3.5 Example of the selection of test lots in research areas of different size (schematic representation, not true to scale; from Bastian, 1991, 1992; design: Kiessling)

Naturalness of vegetation

The intensively used landscapes of Central Europe are dominated by unstable, human-influenced types of vegetation. The stronger the impact of human influence, the more the vegetation differs from its potential natural state (growing on the same sites without human influence). A measure for this difference is the degree of naturalness of the vegetation or hemeroby (Schlüter, 1982, 1984).

Regenerative capacity, age, duration of development

For planning purposes it is very important to distinguish biotopes or eco-systems with regard to their age. A biotope/ecosystem is more vulnerable for nature conservation the less it can be regenerated. For example, a meadow on a former arable field takes one year or less for regeneration, while a beech forest takes up to 100 years. Thus, biotope types can be differentiated with regard to their age/regenerative capacity (Table 3.5).

Diversity

Diversity is a very complex term. It can be related to a number of features: the number of species within an ecosystem; the heterogeneity of an ecosystem mosaic; or the richness of a landscape in biotopes, landscape elements or land-use types. The diversity principle has a common and fundamental significance for the functioning of landscapes, especially concerning matter fluxes, animal migration and aesthetic values, but there is no conclusive relationship between diversity and stability. Ecologically very stable ecosystems exist which are very poor in species.

Table 3.5 Duration of development (age) of several ecosystems (biotope types). From: Bastian, 1994

Age class	Development (Age)	Examples
I	<5 years	short-lived ruderal vegetation, segetal communities, initial stages of rough meadows on sand, vegetation of clear felled areas
II	5–25 years	meadows being poor in species, herbaceous perennial vegetation, ecotone communities, vegetation of eutrophic waters, poor rough meadows on sand, ruderal shrubs and primary woods
III	25–50 years	older (but still little differentiated) hedges and shrubs, oligotrophic silting vegetation, relatively rich reeds, meadows, mesoxerophytic meadows and heaths
IV	50–200 years	relatively rich vegetation of forests, bushes, hedges
V	200–1000 years	fens, transitional bogs, old richly differentiated dry meadows and heaths
VI	1000–10,000 years	peat bogs, old fens, forests with old soil profiles

Spatial criteria

Spatial criteria such as size and arrangement are very important for biotope evaluation. The larger a biotope, the more species can settle there – especially those with a high minimum area demand; also the better the chances are for the maintenance of stable populations, both for genetic and environmental reasons. If the surrounding biotopes differ very much, they have an isolating effect. On the other hand, similar biotopes in the neighbourhood can function as stepping stones or biocorridors.

Representativeness

If a biotope type is representative for a special landscape, it should be kept or developed with priority.

Rarity/degree of threat

In view of the dramatic decline of certain plant and animal species, including their habitats, some species and ecosystems need special attention. However, there is a difference between rarity and degree of threat. There are species and biotopes which, for natural reasons, seldom occur but are not necessarily endangered. In contrast, there are species and biotopes occurring very frequently which are endangered and declining in occurrence.

A number of methodologies are available to combine some of the criteria mentioned above into a so-called *complex biotope value* (representing aggregated indicators; see Table 3.6, Figure 3.6). As an example, Figure 3.7 shows grassland distribution and change in its biotope values (see Table 3.6) between 1950 and 1994. A drastic decrease in biotope values is obvious. This is due to a changing distribution of meadows and arable fields, intensification, drainage, ploughing, and the development of fallow land (set-aside).

Complex biotope values are needed for planning purposes, because they can be handled by the authorities more easily. But the disadvantages of such complex values should not be forgotten:

(1) the information of single parameters are lost;
(2) mathematical processing (formalization) can lead to a false representation of the real situation.

SUMMARY

Although the relevant characteristics of landscapes throughout Europe differ strongly, it is possible and appropriate to apply the same, or at least similar, concepts for the analysis of the natural resource base and for the description of the ecological aspects of landscape change. Thus, different regions can be compared concerning common features, trends and peculiarities. Fundamental

Table 3.6 A possible gradation for the evaluation of biotope types (from Bastian, 1994) (1 - highest... 5 - lowest biotope value)

1 Very endangered and essentially declining biotope types with high sensitivity to anthropogenic impacts and with long time for regeneration; habitat for many rare and endangered species; mostly a high degree of naturalness and only extensive or no use, hardly or not at all replaceable, absolutely worthy of protection

2 Endangered and declining biotope types with a medium sensitivity, with medium to long regenerative times; important as habitat for many, partly endangered species; a high to a medium degree of naturalness, medium or low land-use intensity, only partly replaceable; worthy of protection or improvement

3 Common endangered biotope types with low sensitivity, rather quickly regenerable, as habitats at best of medium importance. A development to more valuable biotopes, but at least maintenance of the present state should be achieved

4 Very common, heavily impacted biotope types, as habitat nearly without significance, low degree of naturalness, short regenerative times, transformation to ecosystems closer to nature is desirable

5 Very heavily impacted, devastated or sealed areas, an improvement of ecological situation is necessary

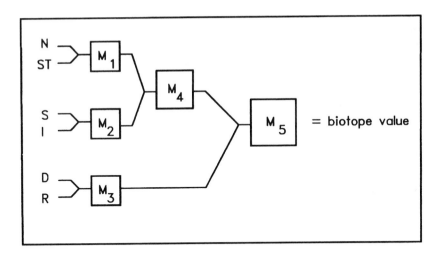

Figure 3.6 Combination of single criteria in an ecological interlacing matrix when evaluating biotopes in the city of Dresden (from Bastian, 1992). N – degree of naturalness of vegetation, ST – structural diversity, S – size, I – isolation, D – duration of development (age), R – rarity, singularity

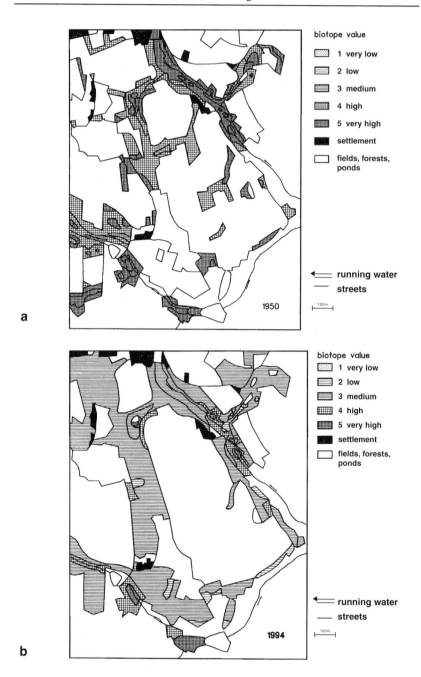

Figure 3.7 Change (mostly decrease) of grassland biotope values between 1950 (3.7a) and 1994 (3.7b) in a test area near Steina (Western Lusatia, Saxony); in 1950 the biotope values 1, 2 and 3 do not occur, in 1994 grade 1 was not found

trends can be recognized with regard to landscape changes in Central Europe. Any methodological approach to the investigation of landscape ecological aspects of landscape change should contain the following main steps: analysis of important natural geofactors and land use at appropriate scales (the indicator principle); diagnosis (evaluation) of natural potentials/landscape functions; the assessment of impacts, resilience and ecological stability; the comparison of different scenarios of future landscape development (landscape prognosis); and landscape management as the sum of all measures to maintain or improve landscape structure and functioning. As an example, the assessment of the biotic regulation potential has been demonstrated.

REFERENCES

Bastian, O. (1987). Grünlandvegetation des Nordwestlausitzer Berg- und Hügellandes einst und jetzt. *Veröff. d. Museums d. Westlausitz, Kamenz*, **11**: 65–82

Bastian, O. (1990). Structure, function and change – three main aspects in investigation of biotic landscape components. *Ekológia* (CSFR), **9** (4): 405–18

Bastian, O. (1991). *Biotische Komponenten in Landschaftsforschung und -planung – Probleme ihrer Erfassung und Bewertung.* Habil. work, M.-Luther-Univ. Halle-Wittenberg, pp. 240

Bastian, O. (1992). Zur Analyse des biotischen Regulationspotentials der Landschaft. *Petermanns Geograph. Mitt* **136** (2+3): 93–108

Bastian, O. (1994). 4.4.2. Regenerationsvermögen und Ersetzbarkeit., 4.4.6. Komplexe Biotopwerte. In Bastian, O., Schreiber, K.-F. (eds) *Analyse und ökologische Bewertung der Landschaft.* G. Fischer Verlag, Jena, Stuttgart, pp. 279–82, 300–22

Bastian, O. and Bernhardt, A. (1993). Anthropogenic landscape changes in Central Europe and the role of bioindication. *Landscape Ecology*, **8** (2): 139–51

Bastian, O. and Haase, G. (1992). Zur Kennzeichnung des biotischen Regulations-potentials im Rahmen von Landschaftsdiagnosen. *Z. Ökologie u. Naturschutz*, **1**: 23–34

Bastian, O. and Sandner, E. (1991). Is an uniform concept for landscape planning imaginable in the future? *LaLUP* **18**: 13–6, University of Massachusetts, Amherst

Bechmann, A. (1989). Bewertungsverfahren – der handlungsbezogene Kern von Umweltverträglichkeitsprüfungen. In Hüber, K.-H., Otto-Zimmermann, K. (eds) *Bewertung der Umweltverträglichkeit,* Taunusstein, pp. 84–103

Buček, A., Lacina, I., Löw, J. and Zimová, E. (1985). Territorial'nye sistemy ekologičskoj stabil'nosti landšafta. *Studia Geographica*, **88**: 136–49

Gigon, A. (1984). Typologie und Erfassung der ökologischen Stabilität und Instabilität mit Beispielen aus Gebirgsökosystemen. *Verhandl. d. Ges. f. Ökologie*, **12**: 13–29

Haase, G. (1978). Zur Ableitung und Kennzeichnung von Naturraumpotentialen. *Petermanns Geograph. Mitt* **122**(2): 113–25

Haase, G. (1990). Approaches to, and methods of landscape diagnosis as a basis of landscape planning and landscape management. *Ekológia* (CSSR), **9**(1): 31–44

Haase, G. (1985). Ansätze und Verfahren der Landschaftsdiagnose als Grundlage von Landschaftsplanung und Landschaftsgestaltung. In Proc. Int. Symp. on *Problems of Landscape Ecol. Research*, 21–26.10.1985, Pezinok (Slovakia)

Haase, G., Barsch, H., Hubrich, H., Mannsfeld, K., Schmidt, R. (1991). Naturraumerkundung und Landnutzung. Geochorologische Verfahren zur Analyse, Kartierung und Bewertung von Naturräumen. *Beiträge zur Geographie*, **34**: 373

Leser, H. (1989). Biotische Faktoren und Bioindikatoren als methodische Probleme landschaftsökologischer Forschungen.-Das 14. "Basler geomethodische Coloquium". *Geomethodica*, **14**: 5–17

Mannsfeld, K., (1979). Die Beurteilung von Naturraumpotentialen als Aufgabe der geographischen Landschaftsforschung. *Petermanns Geograph. Mitt.*, **123**(1): 2–6

Mannsfeld, K. (1985). Landschaftsdiagnose als Beitrag zur Charakterisierung des Landschaftswandels. *Sitz. berichte d. Sächs. Akad. d. Wiss. zu Leipzig, math.-nat. Kl.*, **117**(4): 57–70

Muhar, A. (1995). Plädoyer für einen Blick nach vorne – Was wir aus der Geschichte der Landschaft nicht für die Zukunft lernen können. Symposium *Vision Landschaft 2020*, Eching/München, Mai, 3–5, Laufener Seminarbeiträge 4/1995, pp. 21–30

Neef, E. (1966). Zur Frage des gebietswirtschaftlichen Potentials. *Forsch. u. Fortschr*, **40**(3): 65–70

Neef, E. (1967). Die theoretischen Grundlagen der Landschaftslehre. H. Haack, Gotha, Leipzig, pp. 152

Neumeister, H. (1984). Zur Belastbarkeit und zur Kontrolle von Prozessen und Effekten in der genutzten Landschaft der DDR. *Wiss. Mitt. Inst. f. Geogr. u. Geoökol. AdW d. DDR*, **11**: 7–81

Niemann, E. (1982). Methodik zur Bestimmung der Eignung, Leistung und Belastbarkeit von Landschaftselementen und Landschaftseinheiten. *Wiss. Mitt. Inst. f. Geogr. u. Geoökol. AdW d. DDR*, 2, pp. 84

Schlüter, H. (1982). Geobotanische Kennzeichnung und vegetationsökologische Bewertung von Naturraumeinheiten. *Arch. Nat. Schutz. Landsch. forsch.*, **22**(2): 69–77

Schlüter, H. (1984). Kennzeichnung und Bewertung des Natürlichkeitsgrades der Vegetation *Slov. Acad. Sc. Slovacae Ser. A. Suppl.*, **1**: 277–82

CHAPTER 4

LAND USE, NATURE CONSERVATION AND REGIONAL POLICY IN ALENTEJO, PORTUGAL

F. Bacharel and T. Pinto-Correia

INTRODUCTION

In the European Union (EU) increasing attention is given to the development of a sustainable agriculture that can contribute to the preservation of rural areas, both in terms of landscape and nature conservation (Commission of the European Communities, 1988). Indeed, nature conservation has become an important non-agricultural objective of the Common Agricultural Policy (CAP). In order to maintain a number of nature conservation features which are considered relevant, certain farming practices and land uses are required.

However, how nature conservation objectives can be reached remains an open question. For example, no policy has yet been created at the EU level which integrates aspects of physical planning with nature conservation. Rather, physical planning in rural areas is indirectly incorporated through agricultural, regional and environmental policies, and the possibilities for finance depend on those policies and on national/regional schemes.

In the absence of an EU level integrated policy with a clear financial structure, the combination of EU policy measures with the local planning process might be a way of developing nature conservation. This strategy has been used in the municipality of Castro Verde, with the help of the municipal Master Plan which has recently been approved.

This paper focuses on Alentejo, in southern Portugal. The Portuguese system of land planning and management is described, especially its regional and local components, as well as the main land-use changes that have recently occurred in the region. Based on the case study of the municipality of Castro Verde, the objectives and problems of landscape management and nature conservation are presented, and the options for their integration in planning and implementation are discussed. The solutions adopted and the results achieved in the municipality are also described.

NATURE CONSERVATION AND PLANNING TOOLS IN PORTUGAL

The concerns and objectives of environmental policy in Portugal were established in 1987 through the Environmental Basis Act; this Act also defines a framework for the implementation of a system of laws. The Act is extremely detailed and ambitious, and reflects the sudden importance given to the environmental and nature conservation in political rhetoric at the end of the 1980s.

The Environmental Basis Act contains a list of objectives concerning prevention, equilibrium, participation, unity in management and action, international cooperation, recuperation and responsibility. The most relevant objectives are:

(1) Maintenance or improvement of landscape heterogeneity by monitoring and active management; the criteria to be followed must be defined for each type of landscape, considering both natural and cultural values.
(2) Classification of the landscape into sectors with different functions, including the delimitation of parks, nature reserves and other protected areas, the protection of biotopes outside these areas, and definition of ecological corridors in order to form a *continuum naturale*.
(3) Updating of nature inventories and the definition of techniques and rules for habitat and species protection, nature monitoring and landscape mapping.
(4) Legal protection of flora, fauna and their habitats, with special attention to endemic and rare or threatened species.
(5) Investment in scientific research concerning nature, including promotion of environmental education at all levels.
(6) Implementation of environmental impact studies within all projects and actions that potentially disturb natural conditions.

Unfortunately, to date, only part of the legislation required to meet these objectives has been approved, and even legally approved actions have not been fully implemented.

National level: legislation

Table 4.1 shows the major laws actually approved in Portugal concerning environmental and nature conservation issues, together with their planning and management instruments.

Regional level administration (CCR) and Regional Plans (PROT)

The Regional Coordination Commissions (CCR) are the regional agencies of the Ministry of Planning and Land Management located in the regions. There are five agencies corresponding to the five large regions: north, centre, Lisbon and the Tejo valley, Alentejo and Algarve. Each CCR is divided into three sub-commissions: Land Management, Planning and Development, and Municipal Administration.

The CCR do not have any political function but are the most effective instrument for implementation of physical planning. The are in charge of coordination of national directives while taking account of local problems and concerns. They manage regional policy, trying to preserve local diversity and defend the region in the European context. In principle, the regional approach

Table 4.1   Major Portuguese legislation for land and nature management

Law	Subject	Objectives
D.L.69/90 D.L.176-A/88	Land management	Elaboration of land management plans Definition of the PROT: Regional Land Management Plan Definition of the PDM: Municipal Master Plan Regulation of soil use and its changes (at regional, municipal and local level)
D.L.448/91	Land allotments	Regulation of land allotments in urban areas and areas for urbanisation
D.L.794/76	Soil use	Definition of the minimum unit of cultivation and of the conditions for the separation of land lots
D.L.196/89 D.L.274/92	National Agricultural Reserve (R.A.N.)	Definition of the agricultural soils under protection, and of possible exceptions
D.L.93/90 D.L.316/90 D.L.213/92	National Ecological Reserve (R.E.N.)	Identification of sensitive ecological systems and regulation of their protection
D.L.19/93	Main Act on protected areas	Classification and regulation of areas to be protected, of local, regional or national interest: • National Parks • Nature Parks • Protected landscapes • Nature reserves • Classified sites, groups of sites, and objects
D.L.274A/88 D.L.60/91	Hunting	Establishment of hunting rules, calendar, and list of protected species Definition of the criteria for creating hunting reserves: • National reserves • Social reserves • Associate reserves • Tourist reserves
D.L.316/89 D.L.75/91	Species and habitat protection	Implementation of the Bern Convention Implementation of the EC Birds Directive
D.L.17/77 D.L.172,3,4, 5/88 D.L.139/89 D.L.528/89	Forest	Protection of cork oak and holm oak and of their populations Regulation of woodcutting, of destruction of natural vegetation and soil, and of afforestation with fast growing species
D.L.2/88 D.L.37/91 D.L.468/71	Water	Regulation of the use of water from the public water supply Elaboration of management plans for public dams

Continued

Table 4.1 *Continued*

Law	Subject	Objectives
D.L.70/90		Definition of the public domain concerning use of water reserves
D.L.327/82 D.L.328/86 D.L.149/88	Tourism	Definition of the tourist regions Regulation of the tourist industry
D.L.89/90 D.L.90/90	Raw materials	Regulation of raw materials extraction and of extraction permits
D.R.38/90 D.L.186/90	Environmental impact assessment	Indication of projects requiring EIA Definition of EIA contents

should lead to an integrated understanding of the problems and an overall knowledge of the territory, making it possible to find alternatives and solutions viable in the regional context.

One of the main functions of the CCR is to follow the preparation of Regional Plans (PROT) and Municipal Master Plans (PDM), giving technical advice and helping to define objectives. As legally defined, the main objectives of the Regional Plans are:

(1) Characterisation of the region, its problems and aptitudes.
(2) Regulation of the use of natural resources and the definition of land capability and suitability.
(3) Definition of guidelines for the development of human occupation and land use.
(4) Constitution of a framework for local development and municipal planning.

The Regional and Municipal Master Plans should constitute the framework for land management, but the legal text does not specify which type of region should be addressed in each Plan, nor which agency is responsible for its development. Only a small part of the country is covered by a Regional Plan, and the existing plans do not use a common structure. Regarding their contents, those Plans which have been approved refer to socioeconomic development, infrastructure networks and urban expansion; aspects related to landscape and nature conservation have not been developed.

Municipal level and Master Plans (PDM)

Master Plans (PDM) at a municipal level are considered to be the basic instruments for physical planning; they provide a description of the problems and potentialities of the municipality that result from biophysical, socioeconomic and cultural issues. Furthermore, they constitute the basis for definition of alternative actions required in relation to changes in the national or international (European) contexts.

After 1990, each municipality was obliged to have a Master Plan legally approved by the central administration. Since there is no tradition of local planning in Portugal, many municipalities did not address this legal requirement, justifying their behaviour by the lack of technical expertise and/or economic resources. In view of this situation, the central government ruled that local councils could not apply for European finance without having their Municipal Master Plan approved. As a result, considerable activity has taken place in developing the needed plans but, due to lack of technical expertise within municipal councils, most plans have been completed by private firms. This situation has made supervision by the CCR even more necessary to ensure the quality of the studies and of the final product sold to the local councils.

The legally defined objectives for each PDM are:

(1) To define strategies for socioeconomic development, land use, the urban pattern, and improvement of production activity, infrastructure and equipment.
(2) To furnish local data required for the composition of regional or national plans and apply the guidelines defined in these plans, when they exist.
(3) To constitute a context for the participation of the local population in decisions concerning physical planning.
(4) To classify the territory of the municipality, mainly as urban zones and connections between them, according to their use and future priorities.
(5) To secure adequate use of natural resources, the environment and cultural heritage.

These plans cover the whole territory of the municipality and cover such factors as the aims of development, legal constraints, biophysical character-istics, distribution of economic activities, requirements for urban land, equipment, traffic network and infrastructure. The plans establish:

(1) A spatial framework for the territory.
(2) Land classification, resulting in a spatial zoning which considers urban zones, zones for urbanisation, industrial zones, agricultural areas, forests, areas of natural and cultural uses, and infrastructure.
(3) Rates of urbanisation.

The main concern is urban and infrastructure planning, connected with socioeconomic development; environmental management and the preser-vation of nature and natural resources are secondary objectives. After being approved, the Municipal Master Plans are administered by local Councils. Knowing that most of them have limited technical and economic means, the question posed is if and how they will be able to implement and secure their planned objectives.

CHARACTERISATION OF THE ALENTEJO REGION

The region of Alentejo in Southern Portugal is clearly part of the Mediterranean world and so reflects its most important characteristics and problems. Naveh (1991) estimates that more than 50 percent of the land in the Mediterranean zone has marginal characteristics, due to a combination of limiting environmental factors, structural problems and a lack of regional dynamism. All of these features can be applied to Alentejo.

Biophysical characteristics

Alentejo, occupying approximately one third of the country (Figure 4.1), is the largest Portuguese region, with unique characteristics of land use and land-scape. This region comprises 61.5 percent of lowland (below 200 m) and is dominated by plains, extended fluvial basins and gently folded areas, with only limited mountain areas. Geologically the area is dominated by the Hercinic Massif, with two well defined units: the tertiary basins of the Tagus and Sado rivers, and a large extension of complex schist rocks. The soils are high eroded, very thin and rocky, usually with a low capacity for water retention, poor in organic matter and, therefore, of low fertility.

The climate is typically Mediterranean: the summers are long and dry, with maximum temperatures of 30–40°C; precipitation, irregularly distributed and with considerable annual fluctuations, is between 500 and 650 mm on average, concentrated at the end of autumn and the beginning of winter, with a secondary maximum in March. The rainfall decreases as the temperature increases towards the south and east.

Land use

The distribution of vegetation and the physiognomy of plant associations reflect the conditions of the soil, climate, and the history of human action. In Alentejo, the latter has been responsible for the reduction of woodland and degradation of shrubby communities, but human action has also created, over many generations, balanced agroecosystems adapted to existing environmental restrictions.

Nevertheless, during the last decades human intervention has been insensitive to the great diversity of situations existing in former years. Recently, traditional land management practices have been abandoned, thereby reducing the diversity of agroecosystems and the heterogeneity of the landscape. As in other Mediterranean regions, this has led to an intensification of land use in some areas (i.e. those with the best soils) and an extensification or even abandonment of land in the remaining areas (i.e. those with poor soils, in remote locations, or with hilly or mountainous land) (Pinto-Correia, 1993a).

A number of factors can be considered responsible for these changes: 1) protectionist campaigns for expansion of cereal cultivation, starting in the

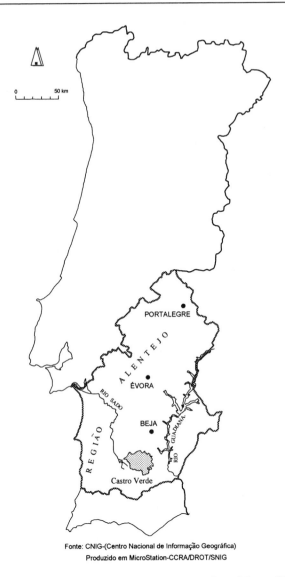

Fonte: CNIG-(Centro Nacional de Informação Geográfica)
Produzido em MicroStation-CCRA/DROT/SNIG

Figure 4.1　Portugal, Alentejo region and the municipality of Castro Verde

1930s and causing an accelerated erosion of thin soils; 2) mechanisation of soil cultivation, beginning in the 1950s and increasing progressively after-wards; 3) application of irrigation, even before the possibilities of improve-ments to non-irrigated land had been exhausted; 4) expansion of rapidly growing tree species (eucalyptus), introduced mainly in the 1970s and 1980s but limited by the fall of world market prices and the application of national legislation; and 5) disintegration of the rural social structure leading to an out-

71

migration to urban centres, or other countries and the depopulation of the countryside. Industry arrived late in this region and has never become important.

Socioeconomic factors

Agriculture has traditionally formed the economic base of most Mediterranean regions and the agricultural changes of the last two decades have led to complex problems for the rural socioeconomic structure. Alentejo is but one of the affected regions: with its natural and socioeconomic fragilities, this region is currently in a process of biophysical and human desertification, deepened by the impact of the structural policies of the CAP.

At the beginning of the 1990s the average population density in Alentejo was 22 inhabitants per km², with some sub-regional variations. The age structure of the rural population reflects the trend of depopulation and shows high ageing coefficients (P > 65/P > 14 × 100 at between 200 and 300) for the period of 1981–1991 (Lourenço, 1993). Selective out-migration is intensified by lack of professional jobs, low technical and cultural level of managers of businesses which restricts innovative capacity, low accessibility provided by road and railway networks, and deficient supply of goods and services (health, education, culture).

THE MUNICIPALITY OF CASTRO VERDE

The municipality of Castro Verde, located in Southern Alentejo (Figure 4.1), has been chosen as a case study for a number of reasons: its bio-physical conditions and socioeconomic pattern are characteristic of the region; its territory provides a very important habitat for the great bustard (*Otis tarda*), justifying special protection; and Castro Verde is one of the few municipalities with an approved Municipal Master Plan.

Biophysical and socioeconomic characteristics

The municipality has an area of 569 km², mostly plains and gentle hills, with heights between 160 and 230 m. There are two physiographic units separated by a main ridge which defines the border between the hydrographic basins of the Sado and the Guadiana rivers. Only 17 percent of the area has soils with a good capability for agriculture, while 83 percent is considered to be without quality for farming.

In 1990 the municipality had 8,350 inhabitants; the density of 15 inhabitants per km² is lower than the regional average. The trend in distribution of population is towards a concentration in the two local centres – Castro Verde and Entradas. The age structure of the population reflects the low birth rate and high proportion of elderly people. These features have been accentuated

72

in the last decade with a decrease in the number of children and young people and an increase in the number of people over 65 years of age. The population between 25 and 45 years are underrepresented, reflecting out-migration of the economically active population.

The concentration of population in the two centres, mainly Castro Verde, is related to decline of the agricultural sector, dynamism of the mining sector (the pyrite reserves of the Neves Corvo mines are amongst the largest in Europe) and activities connected with the building sector, services and commerce. Thus the primary sector, including agriculture and mining, today still occupies 50 percent of the active population, while the secondary sector takes 20 percent, and the tertiary sector 30 percent. The industrial sector consists of small agro-food units and wood/cork producers.

In the last decade there has been a strong effort to develop the infrastructure of the local area and the situation improved greatly between 1981 and 1991. Most households now have access to a central water supply, sewerage system and electricity distribution (Municipio de Castro Verde, 1992). The social and educational infrastructure and distribution reflects the population pattern. Existing institutions for young children are mostly concentrated on Castro Verde and other local centres, with a tendency towards a reduction of their number and/or capacity due to the progressive decrease in the number of children. The same applies to primary schools where the rate of failure in 1989/90 was 25.6 percent. The highest educational level in the municipality is the high school in Castro Verde. There are no special schools. The infrastructure for health and social support of the elderly does not cover their needs.

Evolution of a cultural landscape

Changes in land use in the last decades can be readily analysed by comparing land-use distribution recorded in the Agricultural and Forestry Map, published for the whole country at the beginning of the 1960s, with the present-day situation. In Castro Verde, land-use distribution for 1991 is recorded on the Master Plan (Figure 4.2).

The municipality of Castro Verde is homogeneous as regards physiographic, climatic and soil characteristics, but there is diversity in the landscape and land-use systems. Immediately adjacent to each village is a 'suburban mosaic' of land uses; here, small and medium-sized properties are occupied by orchards, olive groves and kitchen gardens. The limited development of the irrigated area results from limitations in water availability. The most frequently irrigated crops are maize, potatoes and leguminous plants. Olive and fruit production is mostly oriented towards self-consumption and, therefore, these land uses are located close to the settlements.

The wider landscape comprises both open fields under cultivation or grazing and open woodlands of different densities ('montado'). In the former land-use system, cultivated fields are normally used for cereal production

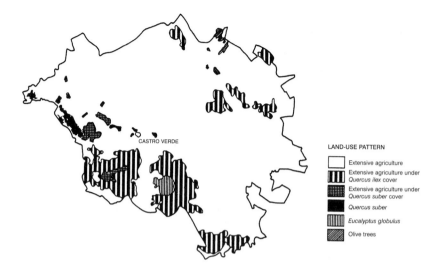

Figure 4.2   Land-use pattern of Castro Verde

(wheat and oats) in a rotation of 3 to 5 years, with 1 to 2 years of fallow. The latter system comprises an agro-woodland pastoral system in which dispersed stands of cork or holm oaks (*Quercus suber* or *Quercus ilex*) are associated with livestock grazing (Pinto-Correia, 1993b).

The riparian vegetation systems also deserve mention: despite their rarity, scattered distribution and poor quality, they are of importance for ecological diversity and biological balance of the local landscapes. These plant associations are essential for protection of the edges of watercourses, control of water currents, and protection against siltation problems. Some important changes occurred in the period between 1957 and 1991:

(1) A decrease in the area occupied by open woods of cork and holm oaks.
(2) Planting of new olive groves and removal of old ones.
(3) Extensification of agricultural practices, mainly transformation of cultivated fields into pastures.
(4) Planting of eucalyptus (600 ha).

In 1991 open oak woodlands in the south of the territory occupied only 12.5 percent of the municipality. These systems are protected by national legislation: they represent rich flora and fauna communities; and have a fundamental role in water resources management, as they promote rainfall infiltration rather than surface runoff and so prevent erosion.

As a result of the changes of the last decades, there are now large plantations of an imported tree species – eucalyptus – in Castro Verde, while the largest part of the municipality is characterised by an open landscape under extensification. The traditional rotation of pasture, crops and fallow has been replaced by extensive grazing, and shrub communities are developing.

Nature conservation

The combination of land-use systems, as currently maintained, supports a large population of small mammals and an extremely important community of birds (reported in 79/409/EEC). Among the bird community, the great bustard (*Otis tarda*) is the most significant, comprising 3.5 percent of the European and up to 60 percent of the Portuguese population. These birds need open fields since their diet is based on cereals. Indeed their ideal habitat is cereal fields combined with pastures, with a low or non-existent application of fertilisers and pesticides, as traditionally practiced in the region. As a result, the munici-pality integrates an area of 88,000 ha constituting an Important Bird Area (IBA), classified as one of the seven priority areas in Portugal for preservation of habitats (Figure 4.3).

According to EC legislation, 'the protected species must be covered by con-servation measures related to their habitat in order to guarantee their survival and reproduction in the area of distribution'. The population of the great

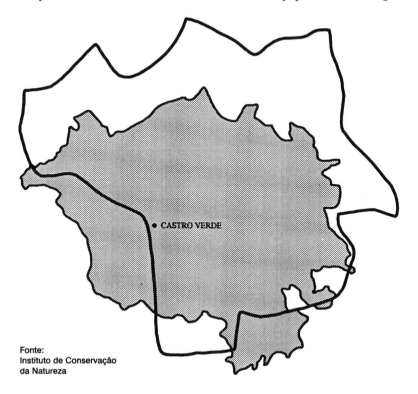

Fonte:
Instituto de Conservação
da Natureza

Figure 4.3   Limits of the important bird area of Castro Verde

bustard is actually decreasing in Europe as a result of changes in agricultural practices of recent years, namely intensification. The preservation of this species justifies maintenance of traditional land-use systems still existing in Castro Verde. If fields are abandoned and shrubs develop further, if previously open areas are planted with eucalyptus, the great bustard will disappear from the area.

IMPLEMENTATION OF PLANNING INSTRUMENTS IN CASTRO VERDE

The municipality of Castro Verde is not covered by any Regional Plan but it has a Municipal Master Plan, approved in 1993 (Municipio de Castro Verde, 1992 and 1993). As in many other municipalities, the plan has been prepared by a private enterprise based in Lisbon, but its development has been closely followed by the Regional Coordination Commission of Alentejo (CCR Alentejo). Discussion of strategies and objectives has taken place with the Municipal Council, resulting in the quality required of the plan.

Strategies and priorities

It has been emphasised in discussions on the Municipal Master Plan, and later specified in the text of the Plan, that one of the priorities of the Municipal Council is to preserve the habitat of the great bustard in the territory of Castro Verde. For this, collaboration of farmers has been sought. The other local objective has been the development of the municipality: creation of new jobs and improvement of quality of life for the rural population. The strategies to be followed attempt to respect traditional life styles and land-use systems, since agriculture has always formed the economic base of the region.

According to the text of the plan, the existence of special nature conservation interests in the municipality should be regarded as potential sources of income. However, appropriate measures need to be implemented to act as the basis for a successful nature conservation policy. Here, the main objective is the preservation of traditional land-use systems, and for this to occur acceptable socioeconomic conditions have to be created for maintenance of the rural population, possibly by supplementing the farmers'/land owners' income.

Solutions adopted

After hearing the views of different interest groups, and having organised public discussions on the problem with farmers/land owners, it was agreed that land use should be managed by zoning the territory into three classes according to the different types of agroforestry management (Figure 4.4):

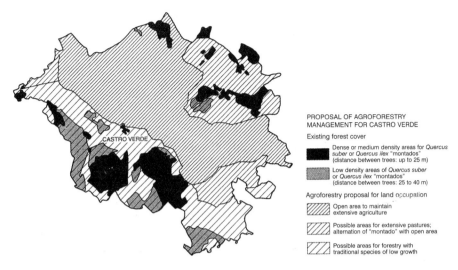

PROPOSAL OF AGROFORESTRY
MANAGEMENT FOR CASTRO VERDE

Existing forest cover

Dense or medium density areas for *Quercus*
suber or *Quercus ilex* "montados"
(distance between trees: up to 25 m)

Low density areas of *Quercus suber*
or *Quercus ilex* "montados"
(distance between trees: 25 to 40 m)

Agroforestry proposal for land occupation

Open area to maintain
extensive agriculture

Possible areas for extensive pastures;
alternation of "montado" with open area

Possible areas for forestry with
traditional species of low growth

Figure 4.4   Proposal for agroforestry management of Castro Verde

(1) Open areas, to be kept in cereal cultivation and extensively used natural pastures.
(2) Areas to be used in extensive woodland–pastoral systems; a mosaic of areas with open tree cover with traditional tree species and open land.
(3) Areas to be eventually planted with forest, only using traditional tree species.

The borders of the different zones were established through discussions amongst those concerned with the plan – the local population and municipal managers. It now lies within the competence of the environmental officials in the municipality to establish detailed rules to be applied in each type of area, as well as priorities to be used in allocation of compensation funds to farmers. In addition, contacts have been established with the Ministry of Agriculture and the Ministry of Environment and a collaboration has been started so as to apply the agroenvironmental measures of CAP to this territory.

Evaluation of current results

Considering, on the one hand, the changes which have recently occurred in Alentejo, not only in land use but in the socioeconomic pattern in general, and on the other hand, the national and European contexts concerning agriculture and land management, the solution adopted is a viable alternative land-use pattern. It combines preservation of the habitat of the great bustard with rekindling of local dynamism required for maintenance of the rural population. It is now up to the farmers to profit in the best way they can from their role as 'nature managers'. They may combine direct income support from 'accompanying measures' of CAP with other income from activities associated with a land-use system that integrates agricultural production and respect for the environment, namely rural tourism. The question is whether they will be able to do so, and if they will get financial assistance to help them. Since both the Municipal Master Plan for Castro Verde and the agro-environmental measures of the CAP have been introduced so recently, it is difficult to evaluate how their combination will result in practice in the municipality considered. But EU agricultural policy has been integrated into local planning and the possibilities have been created.

In several EU reports the expectation is expressed that recreation and tourism can contribute significantly to development of rural areas. For example, in *The Future of Rural Society* (Commission of the European Communities, 1988) three reasons are cited for promotion of 'green' tourism in EU rural areas: to meet the demands of a new category of tourist searching for green spaces and nature; to restore and preserve cultural heritage; to create a source of productivity and new employment in rural areas. However, practical projects or actions oriented towards 'green' tourism in Castro Verde do not yet exist.

CONCLUSION

For a region like Alentejo, the most urgent task for planners and managers seems to be, on the one hand, prevention of marginalisation and, on the other, preservation of natural and socioeconomic resources which have been maintained until the present day. As in other Mediterranean regions, the main challenge is to maintain the balance between the different functions of the landscape while securing an adequate life quality for the rural population. In order to achieve these objectives, application of a management strategy based on conservation objectives and on preservation of specific landscapes may be the way to address EU guidelines for the countryside and the best use of available Community funds.

In Portugal the physical planning necessary to integrate management strategies and nature conservation is organised at a local level. It is also at a local level that problems can be identified and eventual solutions formulated. Even so, the regional level is required to coordinate strategies and distribute technical/scientific support and funds. In Castro Verde the strategies formulated in the Municipal Master Plan respect the idea of integrated management. But for its success the human perspective of landscape management must not be forgotten: natural, social, cultural and economic perspectives must be focused to organise the territory for society.

REFERENCES

Bacharel, F. (1992). The impact of land use changes in the south of Portugal in the context of EC Policies. Proceedings of the Third Seminar of EuroMab Network *Land Use Changes in Europe and their Impact on the Environment.* Poznań

Bacharel, F. (1993). The regional land use management in the context of Land Use Plans of the Municipalities. Proceedings of the Forth Seminar of EuroMab Network *Land Use Changes in Europe and their Impact on the Environment – Comparisons of Landscape Dynamics in European Rural Areas.* Leipzig

Commission of the European Communities (1988). *The Future of Rural Society.* COM (88), 501 The European Commission, Brussels/Luxembourg

Lourenço, J. (1993). Mudança no mundo rural e perspectivas para a agricultura portuguesa. Proceedings of the I Congresso Nacional de Economistas Agricolas, *Que futuro para a agricultura na economia portuguesa.* Associacao Portuguesa de Economia Agraria, Lisboa

Municipio de Castro Verde (1992). *Plano Director Municipal.* Vol. 3, Elementos Anexos. EGF-SAGE, Lisboa, pp. 152

Municipio de Castro Verde (1993). *Plano Director Municipal.* Vol. 1, Elementos Fundamentais. EGF-SAGE, Lisboa, pp. 45

Naveh, Z. (1991). Mediterranean uplands as anthropogenic perturbation-dependent systems and their dynamic conservation management. In Ravera, O. (ed.) *Terrestrial and Aquatic Ecosystems – Perturbation and Recovery.* Ellis Horwood, New York, pp. 545–56

Pinto-Correia, T. (1993a). Land Abandonment: Changes in the Land Use Patterns around the Mediterranean Basin. *Options Méditerranéennes, Etat de l'Agriculture en Méditerranée.* CIHEAM/CCE-DGI, 1(2):97–112

Pinto-Correia, T. (1993b). Threatened landscape in Alentejo, Portugal: the "montado" and other "agro-silvo-pastoral" systems. *Landscape and Urban Planning,* **24**: 43–8

CHAPTER 5

RURAL LAND-USE AND LANDSCAPE DYNAMICS –
ANALYSIS OF 'DRIVING FORCES' IN SPACE AND TIME

J. Brandt, J. Primdahl and A. Reenberg

INTRODUCTION

Land-use changes and landscape patterns are influenced by a wide range of parameters in such a complex way that forecasting becomes a difficult task. Within Danish land-use planning, this issue has been realised again and again over the last three decades. In the 1960s, for example, urban expansion into the countryside exploded due to the economic boom. Physical planning in Denmark at that time was not able to guide the rapid land-use change and, as a new main tool, a strict zoning system was set up to prevent urban sprawl into the countryside. However, the recession of the 1970s almost stopped urban growth pressure on the countryside and even zones reserved for urban growth were often reallocated to rural zones.

In the 1970s, a drastic reduction of the amount of small, uncultivated elements in the agricultural landscape was documented and related to ongoing technological and structural changes within agriculture. A public debate followed and field registrations, monitoring systems and planning measures were developed to counteract the threat. But in fact, already before these measures were put through, the tendencies had changed. In recent years, a general stabilisation of the spontaneous development, unpredictable at the end of the 1970s, has been observed. In the 1980s, much effort was put into forecasting the amount and localisation of the expected marginalization of agricultural land in Denmark. But up to now all prognoses (except those based on interviews with farmers!) have fundamentally failed. Until today, only a few hectares of farmland have been subject to spontaneous marginalization.

Generally speaking, we have to recognise that although it might be possible to predict changes at certain larger spatial and temporal scales, such forecasts cannot be extrapolated to finer levels in time and space. In consequence, any attempt to forecast future land use, at least at a local scale, will be doomed to failure. In spite of such inherent difficulties, important insights into the 'driving forces' of land-use dynamics might be gained during the process of defining an analytical framework (Stomph *et al.*, 1994). In this paper a simple framework for the description of land-use changes will be outlined. Three selected aspects of land-use changes will be presented and used to validate the framework as regards its ability to catch the important driving forces behind some major trends in land-use development of rural Denmark.

AN ANALYTICAL FRAMEWORK

Land use and landscape patterns are often described by parameters such as fragmentation, crop pattern, land-cover composition, and biotope structure (Forman and Godron, 1986; Zonneveld and Forman, 1990; Hobbs and Saunders, 1993; Zonneveld, 1995). Changes in cultivation strategies and reclamation or the marginalization of farmland constitute important dynamic aspects. Rural land-use patterns and dynamics are closely related to the agricultural system. Farm type and farm size are important structural characteristics of this system, constituting the spatial aspects of the human use of the landscape. Thus land-scape pattern dynamics are influenced by a variety of factors such as technology, natural conditions, socioeconomics, public policies and cultural factors, as shown in Figure 5.1.

The development of still more powerful *technologies* has enabled farmers to change the environment radically. Such changes have been reflected in clearly recognisable rural land-use changes. The purposeful application of new technologies, followed by an increasing number of regulations and planning measures, often produces many side-effects of which some may be considered as negative environmental impacts. These side-effects can influence land-use pattern dynamics in an order of magnitude which surpasses all planning intentions.

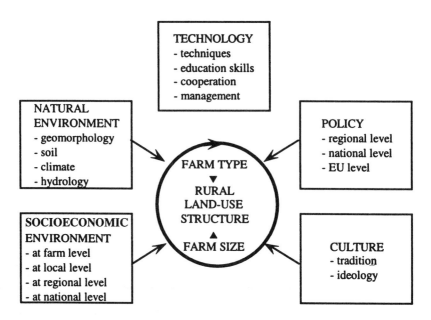

Figure 5.1 Analytical framework. Driving forces influencing landscape pattern dynamics

Variations in natural conditions give rise to considerable regional and local variations in land-use dynamics, and so does a limited understanding of the landscape–ecological conditions and consequences of land-use changes in different landscapes.

Socioeconomic conditions strongly influence land-use pattern dynamics. Prices of input factors and agricultural products, alternative income possibilities and changing economies of scale are all important determinants for land-use strategies at the farm level as well as at the regional, national and global level. Global market developments have led to an increasing division of labour between land-use related industries, such as modern agriculture and forestry, and strengthened the specialisation of land use.

The basic functioning of the capitalist economy should also be considered a core issue for understanding land-use and landscape development (Bowler and Ilbery, 1992). It will nevertheless remain difficult to explain concrete land-use changes on the basis of economic forces only, because of the many state regulations and the mixture of short-term and long-term farmer decisions related to land-use changes in agricultural areas. Rapid changes in *public policies*, especially as regards supra-national levels, have a growing direct and indirect influence on land use (see Bowler and Ilbery, Chapter 7). However, considerable discrepancies in the intentions and the power to implement these decisions at the local and regional levels may occur. These should not be underestimated when the effects of policy on local land-use changes are addressed.

Finally, *cultural* differences and new priorities and ideologies have, under certain circumstances, much more impact on the land-use pattern than normally realised. For example, land-use decisions can often only be satisfyingly explained by incorporating farmers' values.

The *analytical framework* in Figure 5.1 is meant to provide a descriptive analytical framework only; yet, supplemented by relevant statistics and geo-related information, it might be useful as a guide to a profound understanding of land-use pattern dynamics in time and space. The close link between the factors has not been indicated on Figure 5.1, but should be taken into account when using the framework.

LAND-USE DEVELOPMENT IN RURAL DENMARK – AN OVERVIEW

Danish agricultural landscapes have changed significantly within the last two centuries. The landscape patterns and the direction of the changes have been influenced in many cases by structures or characteristics of much older origin. Administrative boundaries (parishes, associations of landowners) have had a long-lasting effect on the basic structure of the landscape, surviving subsequent changes. A profound reallotment reform around 1800, which caused a comprehensive and rational restructuring of the farmland, gave rise to a farm structure with emphasis on middle-sized farms (15–25 ha) of considerable resilience.

As regards general land use – distinguishing land-use classes such as forest, heathland, permanent grassland and arable land – changes can be roughly summarised in an eastern and a western pattern, mainly related to differences in the soil types. The fertile morainic soils in the east have been continuously cultivated and the relative change of importance of the land-use classes is primarily related to draining of wetlands. However, reallotment has influenced the landscape patterns in this part of the country too. As a result of changing property rights for forests and commons, the landscape changed from a relatively atomised field pattern with extended commons, few and open forests and unclear boundaries, into a mosaic landscape with sharp boundaries between intensively used fields, pastures and closed, dense forests, protected for the main purpose of timber production. On the sandy outwash plains in the western part of the country, more radical changes can be observed. Heathland was transformed into cultivated land at the beginning of the century, and a substantial afforestation has taken place on former heathland as well as on abandoned fields.

In a European context, Denmark is generally considered an homogeneous and fertile agricultural region, a characteristic which can be illustrated by the very high proportion of arable land in rotation, even compared with north European standards (Jensen and Reenberg, 1986; Hoggart *et al.*, 1995). Within the last century, a comprehensive inclusion of new land has been made possible by various innovations in agriculture, namely investment in drainage around 1870, planting of hedgerows especially on the sandy soils, and intensive use of marl which has left numerous marl pits as important landscape elements.

The cornerstone of Danish agriculture, from late 1800 to mid-1900, was the mixed family farm; the emphasis of production was on cereals, beet and grass to feed to livestock that provided processed animal products for export (e.g. dairy products). Since 1960, the dominant trend has been towards mechanisation and industrialisation – leading to larger and more specialised farm units. Agricultural policies, planning, and public regulations of various kinds have played an increasing role in rural land-use changes. Today, all types of land-use changes involving agricultural land use would *either* require one or more permissions by the authorities *or* would have to be promoted by one or more subsidy measures.

European Union (EU) policies may currently be the most important single factor in rural processes of transformation. Thus, the 1992 reforms of the Common Agricultural Policy (CAP) had immediate impacts on rural land use. For example, about 8 percent of arable land (210,000 ha) was set aside in the growing season 1992/93 (Andersen *et al.*, 1995: 56). EU agro-environmental policies also have effects on landscape changes through landscape conservation, for example measures related to Environmentally Sensitive Areas (Primdahl and Hansen, 1993).

THE FRAMEWORK IN USE – THREE EXAMPLES

In general, factors which determine landscape changes will be expected to be influential at different spatial scales and with variable strengths. The relative importance of these factors varies considerably with time. In the following examples, three issues mentioned in the introduction – urban fringe landscape development, biotope structure and marginalization – will be used to illustrate analytical opportunities and limitations. Specific attention will be given to the dynamic analysis of spatial patterns in rural landscapes.

Example 1: Rural land use in the urban fringe

Land use in the urban–rural fringe is generally more dynamic than in other rural areas. This is due to a greater number of land-use types occurring in the fringe and to the higher demand for land and consequently a more intensive land market (Bryant *et al.*, 1982). Urban growth and sub-urbanisation are typical land-use changes taking place in fringe areas. In addition, the fringe areas often have several functions occurring on the same piece of land, for example, agricultural production and informal recreational functions. The urban fringe areas, therefore, are often more regulated by physical planning and other types of public control than is the rest of the countryside.

Although land development rights are strongly regulated by the planning system, there is a clear relationship between proximity to cities and land prices in Denmark. For small farms, which, contrary to larger ones, may be sold to anybody interested (farms > 30 ha may only be purchased by buyers with a farming education), land prices vary considerably according to soil quality and proximity to large urban areas. Thus, average prices of agricultural land (including building values) are two to three times higher in areas (munici-palities) near Århus and Copenhagen as compared with the rest of the country. This is an expression of higher demand for land, particularly for smaller farms, rather than speculation on future urban growth. Thus, 64 percent of all farms in the region north of Copenhagen were part-time farms in 1992 (defined as farms operated with a maximum of 0.75 of a man-year), whereas the comparable figure for the whole country was 54 percent (Frederiksborg amt, 1993).

Land use and agriculture in urban fringe areas have been studied only to a limited extent in Denmark. In the following, examples are taken from one of the few studies which have been made in the Greater Copenhagen region.

Land-use structure and dynamics in the Copenhagen region

The location of agricultural land use in eight areas bordering different towns in Greater Copenhagen is shown in Table 5.1 and Figure 5.2. The eight areas are divided between northern locations, where hummocky moraines and terminal moraines with sandy soils dominate, and southern locations

Table 5.1 Agricultural land use and husbandry in eight urban areas in Greater Copenhagen. 1 = Vejby, 2 = Asminderød, 3 = Ganløse, 4 = Smørum Ovre, 5 = Kirke Hyllinge, 6 = Sengeløse, 7 = Tune and 8 = Solrød (from Ogstrup and Primdahl, 1996)

	Northern areas				Southern areas			
	1	*2*	*3*	*4*	*5*	*6*	*7*	*8*
Grain	71	47	55	56	60	50	60	65
Fodder roots	3	0	0	0	5	0	0	0
Grass and green fodder in rotation	5	0	5	0	8	0	0	0
Seeds	6	9	14	22	11	21	31	25
Horticulture	2	0	0	3	5	19	1	2
Permanent grassland	8	19	19	6	7	5	3	4
Set-aside	4	14	6	12	5	4	5	5
Christmas trees and greenery	1	11	0	1	0	2	0	0
100% = (in ha)	241	261	633	383	406	413	383	369
Number of animals per 10 ha agricultural land								
Dairy cows	0.4	0	0.2	0	3.5	0	0	0
Cattle	2.4	0.8	2.2	0.2	8.7	0.1	0.3	0.8
Pigs	10.2	0	1.1	0	45.6	10.6	53.2	7.6
Poultry	7.4	16.4	3.7	0.4	1.2	2.7	2.2	4.3
Sheep	0.1	1.5	1.9	0.5	4.0	0.5	0.2	0.2
Horses	1.0	2.5	0.2	0.4	0.1	0.7	0.3	0.6

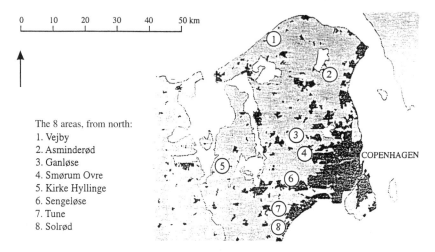

The 8 areas, from north:
1. Vejby
2. Asminderød
3. Ganløse
4. Smørum Ovre
5. Kirke Hyllinge
6. Sengeløse
7. Tune
8. Solrød

Figure 5.2 The eight urban fringe areas studied

dominated by relatively flat and fertile ground moraines. It appears that cereals (mainly wheat and barley) and seeds (rape and seeds for sowing) are the two dominant crops. Forage crops are to some degree grown in three of the areas, and vegetables and horticultural crops are important in three areas including the two nearest to Copenhagen (4 and 6). In three of the four northern areas, permanent grassland is a major land-use category.

Livestock composition varies considerably within the eight areas. In some of the areas intensive pig production occurs, whereas it is absent in two of the northern areas. Except for area 5, located at the furthest distance from Copenhagen, dairy cattle are almost absent. Other grazing livestock (beef cattle, sheep and horses) are relatively abundant in the studied areas. Seen in total, the production and land-use pattern are clearly affected by many part-time and hobby farmers. On the other hand, the Table shows that intensively-farmed holdings exist within the areas, being most significant in the south.

Changes in the ownership structure from 1984 to 1994 are shown in Table 5.2. In five of the six areas for which 1984 data are available, full-time farmers were reduced in number, whereas the number of part-time and hobby farmers grew in most of the areas. Full-time farmers are clearly a minority in all areas. The rate of change within this ten-year period has been quite dramatic. In two of the areas, more than half of the total farmland was farmed by another person in 1994. One of the reasons for this is the great proportion of hobby farmers who lease their land on a one year or short-term basis to the first and/or highest bidder (Ogstrup and Primdahl, 1996).

In some of the areas, the landscape structure has changed dramatically as well. Based on interviews, landscape changes within the last 10 years are shown in Table 5.3. The main tendency is the same as described later: more landscape elements have been established than removed. The rate of change in some areas with respect to new hedgerows and new or restored ponds is remarkable. Again, another striking pattern is the variation between the areas.

Table 5.2 Types of farmers in 1984–1994 in percent. In areas 4 and 6, data from 1984 are not available (from Ogstrup and Primdahl, 1996)

| | Northern areas | | | | Southern areas | | | |
	1	*2*	*3*	*4*	*5*	*6*	*7*	*8*
Full-time[1]	16–5	17–9	9–3	-18	52–32	-14	42–17	20–27
Part-time[2]	16–22	0–10	13–10	-4	9–5	-0	4–0	12–18
Hobby[3]	58–28	52–65	38–45	-43	30–32	-55	25–38	24–23
Pensioners and others	10–45	30–29	40–42	-35	8–31	-31	38–45	44–32
100% = number of owners	19–18	23–21	32–31	-23	23–22	-22	24–24	25–22

[1]Farmers with farm unit as only income source
[2]Farmers with additional income and with more than half of income from the farm
[3]Farmers with less than half of income from the farm

Table 5.3 Relative landscape changes from 1984 to 1994 in eight urban fringe areas in Greater Copenhagen. Hedgerows are shown in m/100 ha, forest and greenery in ha/100 ha, and ponds in numbers/100 ha (from Ogstrup and Primdahl, 1996)

	Northern areas				Southern areas			
	1	*2*	*3*	*4*	*5*	*6*	*7*	*8*
New hedgerows	47	2,400	117	370	233	293	819	98
Hedgerows removed	28	71	6	0	145	0	0	24
New forest	2	0.8	0.2	0.2	0.1	0.1	0.2	0.2
Forest removed	0	0.1	0.2	0.1	0	0	0	0
New greenery	1	8.5	0.2	0.4	0	1.5	0	0
New ponds	0	2	0.6	0.8	0.9	0.7	0.7	0.4
Ponds removed	0	0	0.1	0.2	0	0.7	0	0
Total agricultural area, ha	275	315	807	500	534	461	404	543

Variations in natural environment and farm types cannot explain these differences. There is more variation within the northern areas than between the northern and the southern ones. In an open field landscape like the Tune area 20 km southwest of Copenhagen (Area 7), the new hedgerows will change the present open field landscape dramatically within a few years.

Driving forces in urban fringe land-use changes

Urban fringe farmers have the same *technologies* available as other farmers. However, skills, education and access to technical information are part of the 'technology' concept. There is some evidence that part-time farmers (a significant group in the urban fringe) have lower levels of formal education compared with full-time farmers, while, in the Copenhagen region, relatively few farmers are members of farmers unions: this means that they have less access to advisory services. Consequently, urban fringe farmers tend to operate their farms in a less modern, industrialised way compared with non-urban fringe farmers, although the small size structure of part-time farms also inhibits the adoption of new farm technologies.

The *natural environment* can influence urban form; for example, wetlands and steep slopes are often avoided for urban development and consequently these features tend to form the urban border. This can be seen at different spatial scales in many places in the Greater Copenhagen region. Here, farming is also affected by the lowering of the groundwater table due to groundwater use. The average decrease of the groundwater table in the region north of Copenhagen in this century has been about 5 m, which has made more wetlands available for reclamation; on the other hand, this trend has made it very difficult for farmers to obtain permission to irrigate. Indeed, no permissions to irrigate ordinary crops are granted over most of Zealand because of the competitive use of water for drinking. Finally, urban centres may affect the natural environment by pollution, although this has only been a problem in few places in Denmark.

Socioeconomic conditions include the increasing price level for agricultural land due to demand; this feature has already been mentioned as one of the most significant conditions for urban fringe farming. On the one hand, the higher price level makes it difficult for full-time farmers to maintain an acceptable income from farming; on the other hand, the great number of hobby and part-time farmers provides a relatively large supply of land to lease (rent). Thus, the average farm size (including leased land) for *full-time* farmers is higher in densely-populated regions as compared with the whole country, which is the case in the urban fringe areas studied.

The proximity to urban markets has traditionally been an important locational factor for market gardens, plant nurseries, orchards, etc. Better storage and transportation technologies have reduced the importance of proximity to the market, but it is still a factor affecting land use (Table 5.1).

Planning and other *public regulations* play an important role in rural land use in general and in urban fringe areas in particular. Agricultural and planning policies strongly affect the stability and structure of the fringe areas. The essential aspects of regulations affecting land-use changes are linked to the Danish land zoning system introduced in 1969, which is unique in an international context. It divides the whole country into three zones: (1) rural zone, (2) urban zone, and (3) summerhouse zone. Non-agricultural land use in the rural zone is only allowed on the basis of a so-called 'zoning permit' which is granted or (more often) refused by the regional authorities (Primdahl, 1991). The zoning system is the key to understanding the clear borderline between the urban and rural environment in Denmark. Without such a system, which ensures that all major urban changes occur within the physical planning system and that all projects not related to agriculture are subject to so-called 'rural zone permissions', the Copenhagen area would most likely be affected by urban sprawl. This, however, does not mean that all changes are controlled by regulations.

Suburbanisation in the areas studied is only regulated to a certain extent. For farms smaller than 30 ha, the land market is open; in consequence, many people with urban jobs buy a farm mainly as a rural living place. However, it is not permitted to buy a farm as a second home; the owner must live on the farm. Alternative uses of empty buildings for storage, repair shops, small factories, etc. can take place without a zoning permit. In most areas there has been a varying rate of increase in the number of farms with alternative uses of their buildings – a development which is almost uncontrolled. Furthermore, new farm buildings may be constructed (within some size and height limits) without any permission. The planting of hedgerows and greenery is usually not regulated, apart from some subsidies.

In Denmark there is no specific policy instrument for urban fringe areas, as in the 'green belts' of the U.K. Nevertheless, the zoning system works well in preserving the agricultural component of the landscape. There are no clear signs of underfarmed areas in urban fringe environments in Denmark, which is not the case in many other countries.

On *culture*, urban fringe areas are more dynamic and socially more mixed than other rural areas. This means that local cultural traditions concerning buildings and farming practice may not be easily conserved. Consequently, urban fringe landscapes may often appear more 'untidy' than the rest of the countryside. 'Horsiculture', small industries located in farm buildings, recreational areas like golf courses and similar non-agricultural types of land use are often widespread in the urban fringe. Such conditions may prevent the conservation of local building and landscape management traditions.

Example 2: Small biotopes and landscape dynamics

In the intensively-used Danish agricultural landscape, about one third of the total natural and semi-natural habitat areas for wild plants and animals is made up by so-called 'small biotopes'. This term covers all linear and area elements of less than 2 ha with permanent vegetation or water cover (Agger and Brandt, 1988). Although mainly created by human activity, and commonly viewed as isolated features with little interest from a nature conservation point of view, small biotopes are of importance for landscape ecology because of their stabilising effect, their scenic and recreational functions, and their biological functions as small habitats and dispersal/movement corridors for wild plants and animals.

Small biotopes are an integral part of the agricultural land-use system: most of the 'residual', non-field areas of intensively-used agricultural landscapes are in fact small biotopes. Thus, in many ways changes in small biotopes reflect changes in agricultural land use, although they are subordinate to the factors influencing the agricultural system. As a consequence, small biotopes may be used as indicators for the changes in agricultural as well as other types of rural land use.

Long-term changes of small biotopes

A rough outline of the long-term trend in the development of small biotopes in Denmark is shown in Figure 5.3. Most of these landscape elements are a product of agricultural development. For example, most hedgerows (planted as windbreaks and as enclosures for cattle) and a considerable proportion of the small ponds (mainly resulting from marl pits) were created in the last century in relation to the change towards modern Danish dairy farming. During this process, extensive bogs and wet meadows were drained, giving rise to an increase in open ditches which later disappeared when they were piped.

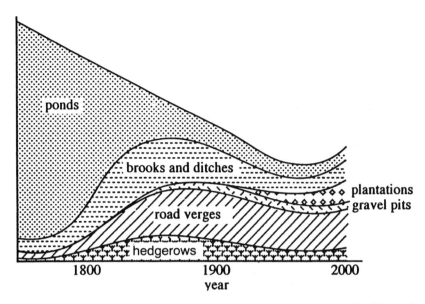

Figure 5.3 A rough outline of the long-term development trend of Danish small biotopes

From the 1960s onwards, the industrialisation of agriculture had been accompanied by a tremendous decline in most types of small biotopes, resulting from the establishment of larger fields on holdings of rapidly growing size and supported by a widespread tendency towards mono-cropping, especially of barley. However, over the last 15 years this trend has been reversed. A stabilisation first observed at the beginning of the 1980s has been followed by a period of increase, which seems to be continuing during the 1990s (Table 5.4).

Trends in the small biotope pattern

It has to be emphasised that the broad trend towards stability in the structure of small biotopes masks diverging regional and local variations. Rapid agricultural specialisation since the 1960s, especially regional variations in specialisation, caused different regional changes in the structure of small biotopes. Also, variations in farm type, as well as in soil conditions, led to different regional tendencies in the dynamics of small biotopes, for instance as observed between eastern Jutland and the islands of eastern Denmark for the period of 1986–1991 (Brandt, 1994).

   In more detail, holdings depending solely on crop production tend to destroy small biotopes so as to obtain bigger and more regular fields as a main way of intensification. In comparison, holdings specialised in livestock production tend to concentrate on technological improvements within farm

91

Table 5.4 The development of small biotopes in Denmark in 1981–91 (Brandt, 1995)

Development of small biotopes in Denmark 1981–1991*		1981–86 (% per year)	1986–91 (% per year)	Wet line biotopes	Wet patch biotopes
13 Test sites	Wet line biotopes	-0.1	-1.1	Drainage ditch	Wet marl pit
in eastern	Wet patch biotopes	-1.8	-0.8	Canal	Other wet pit
Denmark				Brook	Artificial pond
(52 km²)	Dry line biotopes	-0.1	+0.2	River	Bog
	Dry patch biotopes	+0.9	+2.0		Natural lake
					Village pond
10 Test sites	Wet line biotopes		+3.2		Alder swamp
in eastern	Wet patch biotopes		+2.4		Rain water
Jutland					basin
(40 km²)	Dry line biotopes		0.0	*Dry line biotopes*	*Dry patch biotopes*
	Dry patch biotopes		+4.7	Road verge	Dry pit
25 Test sites	Wet line biotopes		+0.3	Field divide	Barrow
in Denmark**	Wet patch biotopes		+0.3	Hedgerow	Plantation
(100 km²)				Slope	Natural thicket
				Railway dyke	
	Dry line biotopes		0.0	Tree row	Solitary tree
	Dry patch biotopes		+2.6	Stone wall	Ruderal area
				Footpath	High power mast

*Indicated as average annual change in percent for all test sites; the line biotopes in percent of length; the patch biotopes in percent of number
**Including two test sites on Bornholm on the Baltic Sea coast

buildings and thereby allow alternative functions for the existing biotope structure. Again, farms situated on good, well-drained soils on flat land tend to continue to eliminate small biotopes, while those on more sandy or mixed soils on hilly terrain tend to stabilise the small biotope pattern (Agger and Brandt, 1987).

The overall pattern of biotopes that in a landscape ecological context can be described as patches and corridors embedded in a matrix of agricultural fields seems to be repeated in a continuous manner throughout Danish landscapes. However, natural conditions are also of great importance for contemporary changes. Today, an increasing number of ponds and lakes are being dug (or re-dug) in former wet hollows which have been drained, while areas of semi-natural vegetation, as well as pastures, have emerged on sandy spots and slopes which are not well suited to modern agricultural machinery.

The removal of small biotopes during the period of agricultural industrialisation was mainly related to the enlargement and regularization of single fields to suit operations with bigger combines and other machinery. Consequently, small biotopes located *within* the land area of single farms were most threatened; whereas biotopes on the *borders* of the holding were more stable. As a result, by 1981 two thirds of the linear and one third of the areas of small biotopes were located at boundaries between two or more holdings (Biotopgruppen, 1986).

Driving forces behind biotopes changes

Both farm size and shape, including the fragmentation of holdings, as well as farm type, including on-going farm specialisation, form basic parameters for the structure and development of the small biotope pattern of agricultural land-scapes. However, behind these basic farm characteristics, a range of interrelated driving forces also characterise and explain the dynamics of small biotopes.

Up to the present day, *technological change* has been one of the most important driving forces. Hedges were adopted as windbreaks and enclosures, especially in the nineteenth century, but the introduction of barbed wire constituted a much cheaper and more flexible way of enclosure that in practice rendered hedges superfluous for enclosure. Marl pits were established in the last part of the nineteenth century, but with the introduction of fertilisers they gradually lost their importance. Later they were used as land-fill sites for the increasing amount of waste created by industry. Today, the land-fill function has been legally prohibited and marl pits are often transformed into game habitats instead.

Ditches were dug as a part of agricultural water management. Later they disappeared because of the introduction of drain pipes, which, until recently, were subsidised by the state. The ditches are, however, now regaining importance because new farm equipment is available; drainage diggers have become a common type of machinery on many farms, and the re-establishment of open ditches, therefore, is again a cheap and efficient alternative to other types of draining.

Variations in natural conditions have given rise to certain regional differences in the pattern and density of small biotopes. A somewhat dense network of hedgerows has been planted on the sandy soils of central and western Jutland. Marl pits were established at the end of the last century on almost every field in the eastern Danish Weichsel morainic landscape, but only occasionally on the sandy outwash plains of western Denmark. Obviously, wet biotopes such as small lakes, ponds, bogs, ditches, etc. are most frequent in areas with loamy soils and a high groundwater table. However, due to the adoption of management practices and other measures that provide favourable conditions for specific types of agriculture, no clear correlation can be observed between variations in natural conditions and spatial variations in the structure and density of small biotopes (Brandt, 1986).

Since the reorientation towards export of dairy products at the end of the nineteenth century, the *socioeconomic conditions* for Danish agriculture have been highly influenced by the cooperative movement. A strong position on the world market for processed livestock products, combined with legislation that favoured middle-sized farms, enabled cooperatives to maintain this size structure of farms until the 1960s, when a concentration and specialisation of holdings started. Although a reduction in the amount of small biotopes has been observed within the last 100 years, the agricultural strategy of the cooperatives has had a stabilising effect by ensuring the profitability of middle-sized holdings. However, the ongoing processes of specialisation and concentration have resulted in more varied socioeconomic conditions for individual holdings and this has weakened the cooperative movement in general and led to a less uniform farm type and size structure. It has been observed, for example, that the density of small biotopes generally drops with increasing farm size. Thus, a diversity in the small biotope pattern is created at a landscape level as a result of the growing differentiation in farm type and size.

With a few exceptions, *public policies* did not directly affect small biotopes before the 1970s. Until then, agriculture was not seen as any threat to nature conservation values. The so-called 'conservation orders' were the only significant conservation instrument for the countryside which, with compensation being paid, could (and still can) protect specific areas (Primdahl, 1991). One exception were the subsidies for planting shelter belts on sandy soils; in the last hundred years, these changed a great part of Jutland from completely open landscapes into a closed-field landscape dominated by a dense pattern of shelter belts.

In recent years, public policies have come to exert an important influence on small biotope dynamics by subsidising drainage, heathland reclamation and other types of intensification. Since 1937 the Nature Conservation Act has contained a 'general protection rule' that defines those landscape elements which must not be removed or actively changed without permission. Refusals are given without compensation and the regulation has been extended several times since it was introduced. Today, moors, heathlands, natural meadows, pastures, and salt marshes larger than 2,500 m^2 are protected by this regulation, as well as all lakes larger than 100 m^2, all barrows, and most of the earth and stone walls in the country (Brandt *et al.*, 1994). This makes public policy a powerful factor in small biotope change. New landscape restoration programmes, as well as new EU agri-environmental policies, are also expected to become important, positive factors in the development of small biotopes (Brandt, 1995).

Finally, *culture* influences the small biotope pattern much more than is normally believed. Several regional peculiarities, such as the impressive lilac hedgerows on southern Funen or the well-preserved stone walls in manorial landscapes, are to be explained culturally. Agro-economic considerations in land-use planning and management can also be subject to culture factors. The growing awareness of the importance of small biotopes obviously contributes to stabilisation of the small biotope pattern. This environmental awareness is linked to contemporary cultural and ideological value changes. Landscape management, including non-productive objectives, is again being seen as part of good agricultural stewardship (Brandt, 1992).

Example 3: Marginalization and landscape structure

In this context, the term *marginalization* will be used for changes in agricultural land towards less economically-intensive uses. Even so, the reclamation of farmland has had, and still has, a significant influence on the landscape pattern of Denmark, as well as in other countries where cultivated land constitutes a major part of the rural landscape.

Statistical trends of marginalization

If we look at the development of cultivated land in Denmark over a period of hundred years, two main regional trends can be distinguished. From 1860 to 1940 there was a gradual increase in the total acreage of cultivated land in the western part of Denmark (Jutland) by an order of magnitude of one third; since then, there has been a modest decline (10 percent). In the eastern parts of the country (dominated by morainic soils), the cultivated area has changed less; the main trend can be summarised as a 10 percent increase until 1880, followed by an equivalent decrease mainly related to urban growth of the cities to the present day.

Recent statistics reveal that these trends continue, but not at the accelerated speed anticipated in the public debate of the mid-1980s; at its peak there was strong public concern about an expected large-scale marginalization of farmland. Nearly one fifth of the farmland was expected to be marginalized before the year 2000. The prognosis caused intensive research related to the marginalization process and to the future use of the abandoned land (Miljøministeriet/ Skov-og Naturstyrelsen, 1987), and influenced the legislation concerning the environment, nature conservation and agriculture (Miljøministeriet/Skov-og Naturstyrelsen, 1992).

Spatial pattern of marginalization

The rough description given above, however, does not suffice to reveal the important spatial patterns in the marginalization process at landscape level. This is illustrated by a closer look at a regional example.

The extent of the afforestation of farmland has been used as an indicator for marginalization on dry soils and to identify the regions dominated by abandoned farmland (Jensen, 1976; Breuning Madsen *et al.*, 1990). The marginalization of farmland has been most intensive on the sandy soils of the central part of Jylland (Figure 5.4). Comparisons with geomorphological maps reveal that the abandoned farmland is located on sandy terraces along the rivers and on coarse sandy soils close to the main limit of the last glaciation. The afforested land, however, is not evenly distributed, as might be interpreted from the map.

The landscape dynamics related to marginalization are rather complex and often include shifts in the various land-use categories (especially cultivated land) which are not visible in traditional statistics (Reenberg and Baudry,

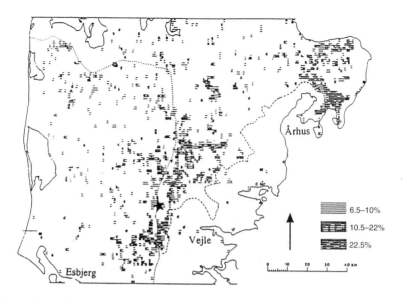

Figure 5.4 The map shows the localisation and intensity of abandoned arable land in a cross section of Jutland. Registrations are made on the basis of a 1-ha grid, in which the percentage of abandoned land is calculated. The intensity is largest on coarse sandy soils in the central part of the area (e.g. on land close to the fertile soils in the eastern part). In the west, where less-fertile soils also dominate, marginalization of farmland has been less well developed because many soils have never been included in rotation in these more remote areas. Asterisk denotes the region represented in Figure 5.5. Based on findings from Jensen (1976)

Chapter 2). Geo-related mapping of land use reveals that landscape patterns have changed significantly and a statistical 'stability' of the cultivated area occurs only because the area of new cultivation (on former meadows) almost matches the area of afforested farmland. The landscape is a highly fragmented and dynamic patchwork of various land-use classes, as shown in the one-year example in Figure 5.5; thus a statistical description fails to provide a satisfactory basis for analysis of the landscape changes.

Driving forces behind changes

Once again, the lessons learned from landscape changes related to the marginalization of farmland can be summarized in relation to the various factors listed in Figure 5.1.

The *natural environment* determines both the enabling and constraining preconditions for cultivation. However, environmental factors, such as soil, geomorphology, climate and hydrology vary substantially in their functions as limiting factors for agriculture because of farmers' *technological* (and thus also economic) ability to cope with natural constraints. The long-term trends

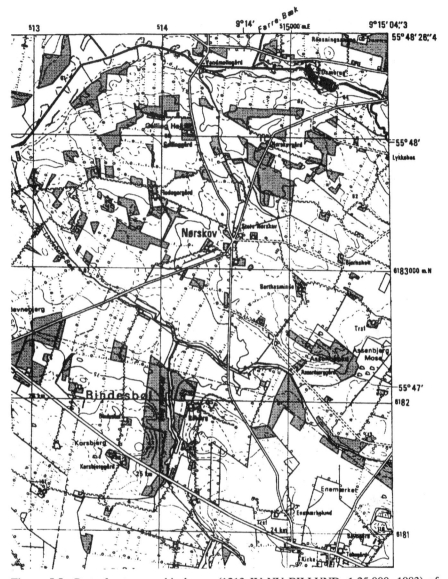

Figure 5.5 Part of a topographical map (1213 IV NV BILLUND, 1:25,000, 1983) of central Jutland. The location is indicated by an asterisk on Figure 5.4. It shows the fragmented landscape pattern dominated by a large amount of more or less randomly distributed plantations on former cultivated land, which is typical for regions dominated by soils marginal to cultivation

reflect, for instance, how the farmers' ability has changed in manipulating water availability by irrigation and drainage. Thus we find that the expansion of cultivated land has largely been located on sandy soils where access to irrigation has been improved, and on organic soils where the majority of wet meadows have been transformed into arable land within the last hundred years. The introduction of fertilisers has reduced the constraints set by low soil fertility and also contributed to this expansion.

The prevailing agricultural 'tradition' (production priorities, crop and livestock composition, etc.) has also played an important role in the value of natural resources. Sandy soils were not marginal in an agricultural system with crop rotations that included grass cover in winter and spring and employed light agricultural equipment. But in the 1960s the shift to spring-grown cereals and larger farm equipment caused marginalization and abandonment. The sandy soils proved too susceptible to wind erosion during heavy storms in the spring when the soil is uncovered by crops. Much of the marginalization can be related to these changes in production traditions at the farm level. Natural factors continuously influence the comparative advantages for farming and are a main condition for the land-use pattern at the regional level.

The *socioeconomic environment* is important at different spatial levels. For example, calculations have shown (Rude and Dubgaard, 1987) that cereal prices, even when reaching rather high levels as in 1985, were approaching the economic limit for profitable cultivation in most parts of Denmark. Farming only slowly responds to economic conditions and, to date, only a small proportion of arable farming on coarse sandy soils has been converted into grassland, afforested or abandoned. Again, land-use strategies in regions dominated by livestock rearing have been influenced by the fact that fodder production practised on the farm, up to now, has proved to be more economic than other feeding strategies.

At the farm level, economics largely determine a farmer's decisions concerning individual fields. For example, farmers can choose not to take into consideration the cost of their own labour at normal rates; an individual farmer might need fields for the distribution of manure in order to fulfil prevailing regulations; or a farmer's debts might be less than average. Consequently, soil maps can be used only partially to identify the fields most likely to be abandoned. For the individual farmer, the production strategy also often varies over the course of time. Thus, the strategies of young farmers are usually more expansionary than those of older ones (Potter *et al.*, 1991). This leads to the patchwork landscapes (Figure 5.5) found in less fertile regions, determined by the decision to abandon the land; such a decision is rooted in the highly diverse economic conditions at the farm level.

National *policies* influence land-use development in general. Examples of specific relevance with respect to marginalization can be found throughout the time period considered, but a general point is that the relative importance of environmental conditions has increased. At the beginning of this century the alternative use of marginal farmland was furthered by public support for afforestation and wetland reclamation. The objective was to develop employment programmes during times of severe unemployment. Later, environmental regulations limited the drainage of potentially acid sulfate soils, started the protection of groundwater resources, and required 2 m of uncultivated 'natural zone' along watercourses. This has led to some reduction of farmland. Other policies, however, have acted as impediments to the marginalization of less fertile soils, as in the case with laws demanding a certain area per animal unit at the farm level.

Recently, the national policy for the regional distribution of land set-aside under the 1992 CAP measures has maintained marginal land ready for use. An equal regional distribution of the amount of land set-aside has been adopted and this strategy has maintained relatively marginal land under cultivation in the less fertile parts of the country.

EU agri-environmental policies have also affected the marginalization process. For example, management agreements have secured the extensive grazing of marginal grasslands which would otherwise have been abandoned. Approximately 50,000 ha of farmland, most of which is permanent grassland, are subject to such agreements (1995) and the number is expected to grow significantly in the years to come. Again, EU beef and sheep premiums are playing a role in preventing, or at least delaying, the marginalization of poor farmland.

Lastly, *cultural* aspects of the marginalization process deserve attention. Marginalization has often been more intense in regions dominated by a relatively early expansion of cultivated land. The example shown in Figure 5.4 reveals that more farmland has been marginalized on poorer soils located close to the fertile morainic soils towards the east. This can be seen as a result of the local perception of appropriate agricultural stewardship. In line with the national spirit in late 1800, and with the examples shown in the fertile regions close by, farmers have tried to include as much land as possible in rotation. Later expansions of farmland onto the less fertile heathlands in the west at the beginning of this century were preceded by a much more critical evaluation of soil quality. In consequence, abandoned land occurs more frequently in the regions cultivated at an early stage of the expansion of farmland.

CONCLUSION

Technology, natural environment, socioeconomics, public policies and cultural values have been suggested as key 'driving forces' in rural land processes. This analytical framework has been applied to three different rural issues: changes in the urban fringe, the dynamics of small biotopes and marginalization. The main conclusion is that the framework is to a high degree useful for the study of changes in rural land-use structures. All major changes are caused by one or more of the five 'driving forces' but the major ones vary in space and time with the specific type of change (see Table 5.5).

Public regulation, including planning policy, together with economics is a major force in explaining changes in the urban fringe. This is especially true for understanding the urban growth process. In the rural part of the fringe

Table 5.5   Driving forces affecting the three types of rural land-use changes

Technology	
A Marginalization	Determine comparative suitability for cultivation of different landscape patches
B Small biotope density	Replace functions needed for biotope creation and management
	Make it easier to remove/establish new biotopes
C Land-use changes in the urban fringe	(NO FACTOR)
Natural environment	
A Marginalization	Determine basic precondition
B Small biotope density	Enable or constrain removal and construction of small biotopes
	Influence type of agricultural system
C Urban fringe changes	Enable and constrain alternative uses, including urban
Socioeconomic	
A Marginalization	Determine limits to feasible agricultural cultivation patterns
	Influence demand for land
B Small biotope density	Lead to specialisation and concentration in agriculture which results in differentiation of small biotope patterns
C Urban fringe changes	Proximity to urban areas leads to high prices and diverse land-use structures
Policy	
A Marginalization	Stimulate desirable changes from arable to grasslands/forests
	Reduce unwanted abandonment of grasslands
B Small biotope density	Stimulate new biotopes
	Protect existing biotopes
C Urban fringe changes	Reduce land speculation and thus prevent 'under-farming'
	Reduce urban sprawl
Culture	
A Marginalization	Determine farmers perception of marginality
B Small biotope density	Determine farmers appreciation of small biotopes
C Urban fringe changes	Determine differences in urban and rural values

areas, natural conditions, economics, public policies and cultural values jointly affect market conditions, ownership structure and the resulting changes of the rural land-use structure. Thus, the analytical framework is a useful tool for analysing the differences between the changes of rural land use in the urban fringe and other rural areas, but it is not very helpful in preparing urban fringe studies in detail.

Small biotopes are also affected by all five 'dynamic forces'. A factor like technology is particularly important in explaining changes over the course of time because it replaces or introduces agricultural functions resulting in the removal or creation of small biotopes. New technologies often have different, sometimes adverse, effects on areas with different natural conditions. Farmers' values related to small biotopes, together with public policies, also play an important role for dynamics.

In describing the marginalization of agricultural areas, the framework is clearly operative. All five 'forces' are of relevance and affect marginalization. Socioeconomic conditions are of high importance for the marginalization process in general, whereas the relationship between technology and the natural environment may have major effects in a specific space–time situation.

In brief, the general conclusion from the material presented is that a multidisciplinary analytical framework is both necessary and useful for the investigation of land-use dynamics. At a general level, it is possible to list the relevant key parameters which should be included in the framework. The relative importance of these parameters does, however, vary considerably and in an unpredictable way in time and space. Consequently, there is no straightforward, if any, possibility of modelling or forecasting land-use dynamics.

However, a framework, such as the one proposed, might prove to be a most valuable checklist in the initial phase of research design. It will be a relevant tool to ensure that important parameters are not left out – an issue which still deserves attention even if it is presupposed that it might not be possible to create a total, quantitative model.

REFERENCES

Agger, P. and Brandt, J. (1987). *Smaabiotoper og marginaljorder.* Miljøministeriets projektundersogelser 1986. Teknikerrapport nr. 35. Miljøministeriet/Skov-og Naturstyrelsen, København

Agger, P. and Brandt, J. (1988). Dynamics of small biotopes in Danish agricultural landscapes. *Landscape Ecology,* **2**: 227–40

Andersen, E., Kristensen, K. and Primdahl, J. (1995). Agricultural marginalisation and land-use regulation in Denmark. In Bethe, F. and Bolsius, E. C. A. (eds) *Marginalisation of agricultural land in The Netherlands, Denmark and Germany.* Ministry of Housing, Spatial Planning and the Environment, The Hague

Biotopgruppen (Agger, P., Brandt, J., Byrnak, E., Jensen, S. M. and Ursin) (1986). *Udviklingen i agerlandets smabiotoper i Ost-Danmark.* Forskningsrapport nr. 48 fra Institut for Geografi, Samfundsanalyse og Datalogi, Roskilde University

Bowler, I. and Ilbery, B. (1992). A conceptual framework for researching the environmental impacts of agricultural industrialization. In Baudry, J., Burel, F. and Hawrylenko, V. (ed.) *Comparisons of landscape pattern dynamics in European rural areas.* EUROMAB Research Programme, UNESCO, Paris, pp. 288–97

Brandt, J. (1986). Small biotope structure as a synthesizing feature in agricultural landscape. In Richter, H. and Schönfelder, G. (eds) *Landscape synthesis – foundations, classification and management.* Martin-Luther-Universitat Halle-Wittenberg, Germany, Vol. I. pp. 52–61

Brandt, J. (1992). Land use, landscape structure and the dynamics of habitat networks in Danish agricultural landscapes. In Baudry, J., Burel, F. and Hawrylenke, V. (eds) *EUROMAB: Comparisons of landscape pattern dynamics in European rural areas.* Vol. 1, 1991, UNESCO, Paris, pp. 213–29

Brandt, J. (1994). Smaabiotopernes udvikling i 1980erne og deres fremtidige status i det åbne land. In Brandt, J. and Primdahl, J. (eds) *Marginaljorder og landskabet–marginaliseringsdebatten 10 år efter.* Forskningsserien nr. 6/1994. Forskningcentret for Skov and Landskab, Lyngby, pp. 21–50

Brandt, J. (1995). Ecological networks in Danish planning. *Landschap,* **12**(3): 63–76. Special Issue on Ecological Networks in Europe

Brandt, J., Holmes, E. and Larsen, D. (1994). Monitoring 'small biotopes'. In Klijn, F. (ed.) *Ecosystem classification for environmental management.* Kluwer Academic Publishers, Dordrecht, pp. 251–74

Breuning-Madsen, H., Reenberg, A. and Holst, K. (1990). Mapping potentially marginal land in Denmark. *Soil Use and Management,* **6**(3): 114–20

Bryant, C. R., Russwurm, L. H. and McLellan, A. G. (1982). *The city's countryside. Land and its management in the rural–urban fringe.* Longman, London

Forman, R. T. T. and Godron, M. (1986). *Landscape Ecology.* John Wiley and Sons, New York

Frederiksborg amt (1993). *Landbrugsundersøgelse* 1993. Frederiksborg amt, Hillerød

Jensen, K. M. (1976). Abandoned Agricultural Land (In Danish with English summary). *Atlas over Danmark II, 1.* Reizel, Copenhagen

Jensen, K. M. and Reenberg, A. (1986). *Landbrugsatlas Danmark.* Reizel, Copenhagen pp. 120

Hobbs, R. J. and Saunders, D. A. (1993). *Reintegrating Fragmented Landscapes. Towards Sustainable Production and Nature Conservation.* Springer Verlag, Berlin

Hoggart, K., Huller, H. and Black, R. (1995). Rural Europe, Identity and Change. Arnold, London

Miljøministeriet/Skov-og Naturstyrelsen (1987). *Marginaljorder og miljøinteresser-en sammenfatning.* Copenhagen

Miljøministeriet/Skov-og Naturstyrelsen (1992). *Naturbeskyttelsesloven.* Copenhagen

Ogstrup, S. and Primdahl, J. (1996). *BynÆre landbrugsomrÂder i hovedstadsregionen 1994.* Forskningsserien nr. 14-1996. Forskningscentret for Skov and Landskab, Hørsholm

Potter, C., Burnham, P., Edwards, A., Gasson, R. and Green, B. (1991). *The diversion of land. Conservation in a period of farming contraction.* Routledge, London

Primdahl, J. (1991). Countryside planning. In Hansen, P. E. and Jørgensen, S. E. (eds) *Introduction to Environmental Management.* Elsevier, Amsterdam

Primdahl, J. and Hansen, B. (1993). Agriculture in Environmentally Sensitive Areas: Implementing the ESA Measure in Denmark. *Journal of Environmental Planning and Management,* **36**(2)

Rude, S. and Dubgaard, A. (1987). Economic investigations of extensive and intensive uses of dry marginal land (In Danish). *Marginaljorder og miljøinteresser,* teknikerrapport 15. Skov-og Naturstyrelsen, Copenhagen

Stromph, T. J., Fresco, L. O. and van Keulen, H. (1994). Land-Use System Evaluation: Concepts and Methodology. *Agricultural Systems,* **44**: 243–55

Zonneveld, I. S. and Forman, R. T. T. (eds) (1990). *Changing Landscapes: an ecological perspective.* Springer Verlag, New York

Zonneveld, I. S. (1995). *Land Ecology.* SPB Publishing, Wageningen

CHAPTER 6

DRIVING FACTORS OF LAND-USE DIVERSITY AND LANDSCAPE PATTERNS AT MULTIPLE SCALES – A CASE STUDY IN NORMANDY, FRANCE

J. Baudry, C. Laurent, C. Thenail, D. Denis and F. Burel

INTRODUCTION

This chapter is based on intensive interdisciplinary research on land-use changes and their ecological consequences in central Normandy. Based on the results, we propose one possible way to carry out research and give an illustration of the general framework proposed by Reenberg and Baudry, Chapter 2. We suggest that the debates are confused and predictions misleading because the numerous levels of social, technical and ecological organization involved are usually not clarified. Extrapolation of the results of ecological research at field scale with no consideration of actual landscape composition and pattern, as well as scenarios of regional changes not considering social and technical factors are among the main areas of confusion. Debates over scenarios of change (Schoute *et al.*, 1995) often stem from this kind of confusion.

Land-use patterns and dynamics are driven by a variety of factors, from farmers' daily activities (ploughing, harvesting) to national and international agricultural (Common Agricultural Policy of the European Union – CAP), social and trade (General Agreement on Tariffs and Trade – GATT) policies. The effects of these span from field to regional scales. Data needed to understand patterns are collected by interviews, drawn from statistics or remote sensing. Several questions must be addressed: 1) How are decisions and patterns at different scales related? 2) How do policy decisions flow from upper to lower levels, especially local dynamics, family and farm structure? It is striking that, although there has been a common, uniform agricultural policy in the European Union for thirty years, farming systems and land-use patterns are still very diverse at any scale (Laurent and Bowler, 1997).

The interdisciplinary research was carried out in the Pays d'Auge in Lower Normandy, France, using methods that:

(1) can be applied at different scales to study spatial patterns;
(2) link patterns from scale to scale; and
(3) link ecological patterns to agricultural dynamics.

CONCEPTUAL FRAMEWORK

The conceptual framework addresses the key concepts used as well as definition of entities within the different disciplines and their relationships and dynamics.

Three key concepts are at the basis of the framework of analysis: landscape, farm technical systems, household systems of activity.

The *concept of landscape* (Zonneveld and Forman, 1989) considers the continuity of space and subsequently pays attention to two points:

(1) How frequent are the phenomena we observe in certain spatial structures or field types?
(2) How do landscape patterns control ecological patterns at the field scale?

The *farm technical system* (Brossier *et al.*, 1993) is the key factor in how farmers use the land, techniques are employed, and alternatives exist for a given system of production (e.g. dairy cattle may be fed predominantly from permanent grassland or from annual crops). The various technical systems yield a particular landscape structure and distribution of input/output in space and time. For a given type of production, changes may be sudden and widespread, creating new land uses and environmental conditions.

The *household system of activity* allows understanding of how farming systems depend on the socioeconomic context. The household is the socio-economic entity within which decisions on the farm are made. The effects are through the diversity of incomes and possibilities to diversify activities. The approach of the household, rather than the farm level, take into account other aspects of the economy than agriculture for its own sake. Farming activities can be linked to social differentiation (Laurent, 1994).

Hierarchical approaches

To handle the complexity of landscape dynamics, entities are used that correspond to levels of organization, either spatial levels (corresponding to mappable units) such as field, farm, landscape, municipality, agricultural or administrative region, or functional levels (used to study processes) such as fields, farms, groups of farms, landscape units.

The paradigm of *hierarchy theory*, widely applied to the study of ecological systems, is that organizational levels which differ by their structure have a certain functional autonomy such that higher levels are more than the sum of the parts and have, generally, a lower rate of behaviour (Allen and Hoekstra, 1992). Because they are almost autonomous, these entities can be analyzed separately and their interactions detailed. The theory emphasizes the importance of the role of the extent of observations in space and time on results and the scale dependence in the perception of phenomena.

Within the economy, *the regulation school* approaches (Boyer, 1986) rely on the assumption that, in the capitalist mode of production, the expanded social mode is always partial, temporary and unstable. Accumulation of capital and maintenance of an appropriate rate of profits imply that the economic system is regulated by different mechanisms. Not only state and market regulations must be studied, but other mechanisms that are interrelated (institutions, social norms etc.). At some stages of the development of capitalism the different mechanisms regulating the economy are coherent enough to ensure economic growth; but when the inherent contradictions of the system become too strong, there is a crisis. These approaches share several commonalities:

(1) Regional landscapes are the product of human activities interacting with the physical environment.
(2) The landscapes are continuous in space and time.
(3) Patterns and processes are studied within abstract ecological and economic spaces constructed by researchers.
(4) They seek to understand dynamic processes and do not assume any type of equilibrium or end-point for the dynamics.

Ecology, economy and agronomy have their own objectives that can be levels of organization. A crucial point in the research is that each of these be studied by at least two disciplines to permit integration. We choose to maintain single disciplinary approaches because we consider that each discipline creates its own theories that can produce different mechanisms. For example, ecology theory has little to say on the way in which farmers choose the techniques they use, though ecology can assess the consequences of different choices on the environment. Conversely, the concepts devised in the realm of agronomy are of little use to analyze ecological mechanisms.

Data are analysed first, to assess the diversity of the different sets of entities (farm, household, etc), then to find relationships within diversity and finally, to test if they are consistent with the functional models used in the different disciplines.

The functional models

Landscape as an ecological system: the central point is that the spatial structure of land utilization and associated field boundaries (hedgerows) governs the distribution of species over an area (Baudry, 1991). In the case of colonization subsequent to abandonment, mature elements (woodlots, hedgerows) would provide seeds and dispersers to newly available habitats. Therefore, depending upon the environment of an ungrazed or abandoned field, the species able to colonize would differ.

The farm as a system (Brossier *et al.*, 1993): the farm family sets objectives of production (type and volume) depending upon its needs, the economic and technical environment (market, machines, labour), and the farm structure (area, physical conditions). To reach the objectives the farmer makes decisions on field utilization that affect all other fields of the farm. For example, to feed a herd of dairy cows a farmer will need a given quantity of grass during the grazing period and fodder as silage or hay during the winter. This limits the area available for other crops (Thenail and Baudry, 1996). Then the assignment of crops to fields will depend upon a variety of factors, from field size and shape to its distance from the farm building. The functioning of adjacent farms in an area will produce the landscape pattern.

THE POINT OF INTEREST: RISK OF LAND ABANDONMENT

In 1983, when milk quotas were introduced in the European Community due to a large surplus in production, many calculations of land surplus were made (Moati, 1989). It was inferred from these calculations that land surplus would be abandoned. For all of France, 6 million hectares were expected to be abandoned by the year 2000; i.e. around 20 percent of the farmland. In Lower Normandy, 300,000 ha were at risk. Our original question in 1988 was 'where are the zones at risk and how shall we manage them?'

The research region was the Pays d'Auge, a formerly wealthy agricultural region renown for its cheese (Camembert de Normandy, Pont-Lévêque and Livarot) and Calvados (apple brandy). It is a sedimentary basin carved by relatively deep valleys oriented from north (seashore) to south. The altitude varies from O to 250 m. It has a rolling landscape with a great variety of substrates from alluvium to limestone and clay on the same slope, hence a diversity of soils and physical constraints. The main constraints are slope and hydromorphy (springs scattered at the limit between clay and limestone). Around 40 percent of farmland is non-arable. As the Pays d'Auge is close to the sea (south of La Manche), the climate is temperate. Monthly mean temperatures vary from 4 to 16°C. Rainfall varies from 37 to nearly 80 mm per month; the average total per annum is around 600 mm.

Permanent grassland covered 87 percent of the farmland in 1988 (General Agriculture Census) and had decreased slightly since 1979 (91 percent). At the end of the seventeenth century and during the eighteenth century fields formerly used for cereal production were laid to grass. At this time cattle meat for the Paris area was the main area of production. At the end of the nineteenth century the introduction of railways produced competition with other regions (The Charolais). Production then shifted to dairy cows. Farms are of small and middle size: 51 percent are less than 20 ha.

Because of slope or hydromorphy, it was considered one of the most threatened regions in France. A casual traveller in the region would see ungrazed patches in grassland, an 'obvious' sign of the start of abandonment.

METHODS OF OBSERVATION

The methods of observation are diverse because various disciplines and objects of study are involved. To insure consistency of observations has been a major point of the research.

A landscape approach requires spatial continuity in observations, which can easily be done for land cover. This is more difficult for farming systems for several reasons: if farm territories are scattered it may not be possible to know all the farmers; also, farmers may not be willing to be interviewed. To sample landscapes and the economic context at the same time, we chose eight munici- palities differing by the type of terrain (topography and hydromorphy) and the vicinity of small towns. The variety of terrain sets the context in terms of physical constraints (climate is assumed to be constant over the study area), and vicinity of towns sets the economic context, including the possibility of out of farm jobs. We interviewed 190 farmers (household with farming activity) who farm 9183 ha, including 4656 ha in the municipalities. As fields are scattered, many farmers have fields in one municipality and their farmstead elsewhere. We obtained information on about 2000 fields in the sampled municipalities.

The research, carried out between 1987 and 1992, showed that land abandon- ment was still more a social fantasy than a real phenomenon, in France it only occurred in limited areas and it is, therefore, worthwhile to examine the reasons for this.

Sampling of species can only be done in a limited number of fields and/or hedgerows. Extrapolation to the entire landscape requires a model relating species distribution to both farming practices and landscape structure. Until now, a complete model has not been made. Plants, carabids and spiders were studied (Burel and Baudry, 1995), as they represent different potentials of dispersal. We sampled over 200 grassland vegetation plots, 34 hedgerows and 51 ungrazed patches.

FARM AND FIELD SURVEYS FOR ANALYSIS AT THE LANDSCAPE LEVEL

The central point is the diversity of farms in terms of land use and how this relates to the diversity of households.

Diversity of farms – land use

For the definition of land-use pattern types, the farms were characterized by four variables: total farmland (excluding woodlots and non-productive land), the portion of farmland under permanent grassland, the portion of farmland under cultivated forage crops (mainly maize for silage), and the portion of farmland under cash crop (mainly cereals). Data were analysed using corres- pondence analysis then cluster analysis to yield groups of farms. These groups were labelled 'Land-Use Systems' (LUS).

Six groups were identified. The importance of grassland diminished markedly from type A to type E (100% to 35% of the total area of the groups) (Table 6.1). Meanwhile, forage crops (from 0 to 39%), then cash crops (from 0 to 26%) increase. Farm size, also an active variable in the analysis, does not discriminate the types as clearly. All farm sizes are represented in type A (Figure 6.1). The largest farm with only permanent grassland is 110 ha.

Table 6.1 Importance of different types of farm as defined by land-use patterns and type characteristics according to the variables used in the multivariate analysis

Farm type (%)	A	B	C	D	E
Farms	57	16	15	6	4
Permanent grassland	100	85	70	47	35
Forage crops	0	11	25	26	39
Cash crops	0	3	6	26	26

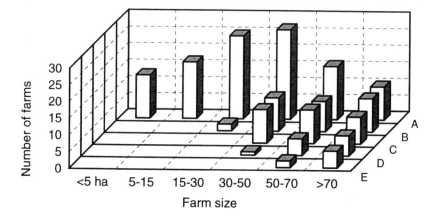

Figure 6.1 Farm size distribution among land-use types A to E (see Table 6.1)

Furthermore, all size classes are more frequent in this type than in any other type. Nevertheless, going from type A to E, larger classes dominate increasingly. There is only one farm of less than 30 ha with ploughed land. Thus farm size is a variable partly correlated with land-use patterns.

The different types do not equally share the territory (Table 6.2). Type A is dominant, while D and E, where ploughing is important, cover only one fifth of the study area. Table 6.2 shows that the different land-use types of farms share the total study area and the investigated municipalities in the same way. There is no bias introduced by large farms, which would alter the land distribution categories.

Diversity of households and integration of farms in the regional context

The differentiation among farms is correlated with a number of social characteristics (Table 6.3). Older farmers are found in type A farms. Farmers in types A and B have the less professional contacts. Farm types are also dependent on the type of household they are part of, especially the structure of family income. Farmers of types D and E, the most intensive in terms of

Table 6.2 Share of territory by different farms' land-use types

Farm type (%)	A	B	C	D	E
Total study area	39	23	20	11	6
Municipalities	36	21	23	13	7
Permanent grassland	49	24	18	6	3
Forage crops	0	19	40	22	19
Cash crops	0	12	19	44	24

Table 6.3 The social context of different types of farms classified according to land use (% type)

Farm type	A	B	C	D	E
a) Farmer's age					
<35 years	10	29	18	58	29
35–50	21	29	39	25	43
50–60	40	29	39	8	29
>60 years	29	13	4	8	0
b) Professional relationships, CUMA : Cooperative for purchase and use of farm machinery					
CUMA	9	52	82	58	86
Extension services	5	35	46	50	57
c) Sources of income					
Society	11	0	0	0	0
Hobby farm	3	3	0	0	0
Part-time farm	19	3	4	0	0
Pluriactivity + pension	6	0	0	8	14
Pluriactivity, no pension	13	16	18	42	43
Professional + pension	16	26	7	8	0
Professional, no pension + direct marketing	7	19	21	17	0
Professional only	24	32	50	25	43

land utilization, are either 'professional' or in a family where someone has a job outside the farm, which is rare in the other groups. Part-time farmers and societies (horse rearing) have farms with only grassland.

STUDY OF FARM TECHNICAL FUNCTIONING TO UNDERSTAND THE MECHANISMS OF DIFFERENTIATION

If land-use and management practices within farms are diverse, this is because it is necessary to attain objectives of production. Each field has a function within the technical system (Thenail, 1992). A detailed analysis of fifteen farms has permitted definition of their production system and associated technical system (means of production). This typology has been extrapolated to most farms of the region. Within each farm type, either grassland or crops can be assigned a productive function associated with specific management practices (e.g. fertilization, mowing, grazing management). In several farms none of the grassland receives fertilizer, but some pasture receives more care (mowing of ungrazed patches) than others.

Types of farm according to function

We defined eight types of farms according to their production system and their type and level of production, taking only stock farms into account. We made distinctions between dairy and meat farms. The most discriminating factors are presented in Table 6.4 and were used to extrapolate the typology to the whole population of farms.

Figure 6.2 shows the relationship between the types of farms defined by land use (LUS) and the types of farms defined according to functioning. Land-use type E has little to do with stock farms. There is a link between land cover and farm functioning but, as expected, a type of land cover can correspond to several types of functioning. This is especially true with grassland dominated systems.

We then considered the physical environment within the farms sampled through characteristics perceived by farmers as possible constraints: slope, humidity, accessibility to parcels (parcels can be very scattered in the region). The relative contribution of variables of physical environment and of the type of farms to explain grassland use was examined. The results show that the type of farm is the most important factor determining grassland use. All farmers of

Table 6.4   Main characteristics of types of farms classified according to functioning

Dairy and dairy + meat farms	Agricultural area (ha)	Milk quotas (thousand liters)	% permanent grassland/AA	Quota per cow	Dairy cows/ cattle units
Dairy+meat 1	10–45	20–90	≥ 90	≤ 4,000	variable
Dairy+meat 2	50–90	50–120	≥ 90	≤4,500	20–60
Dairy+meat 3	50–150	100–300	45–90	variable	30–65
Dairy 1	30–60	80–150	70–100	≥ 4,000	> 40
Dairy 2	40–80	150–300	≤ 70	≥ 4,500	≥ 45

Meat farms	# Cattle units	# Suckling cows + beef cattle	# Fattening beef cattle	#Suckling cows	Forage area (ha)	Stocking rate (cattle unit/ha forage)
Meat 1	< 25	≤ 20	variable	variable	< 36	variable
Meat 2	≥ 25	> 20	0–30	15–40	≥ 36	0.8–1.6
Meat 3	≥ 25	> 20	10–50	0–20	> 36	0.5–1.9

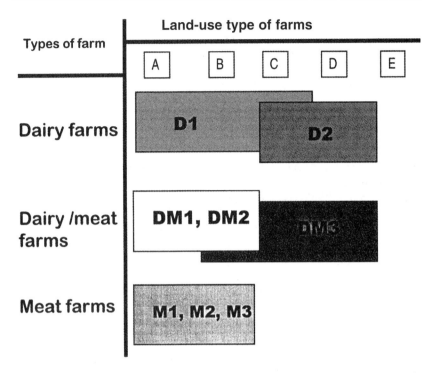

Figure 6.2 Land-use types (see Table 6.1) and types of farms according to their function (see Table 6.4)

the sample have about the same proportion of 'constraining parcels' (30 to 40 percent in area) but do not manage these grasslands in the same way. Management depends essentially on their choice of production system (Baudry *et al.*, 1994).

Diversity of grassland management practices at field scale as related to farm diversity

We tested the relationships between the level of nitrogen fertiliser and other practices like harvesting and animal grazing, using a multivariate analysis on farming practices in grassland. The first ordination axis provides a gradient from extensive (no input) to intensive (more than 150 kg N per annum).

A classification defines five types of grassland use (TGU) that can be characterized in terms of input (fertilization), use (hay, silage) and other management practices (winter grazing). From type 1 to type 5, organic and mineral fertilization as well as silage increase while winter grazing decreases (Figure 6.3). It is worth noting that 50 percent of the parcels under permanent grassland receive less than 50 kg of nitrogen per annum. Many of these grasslands show signs of low grazing pressure, such as patches of brambles. This may be

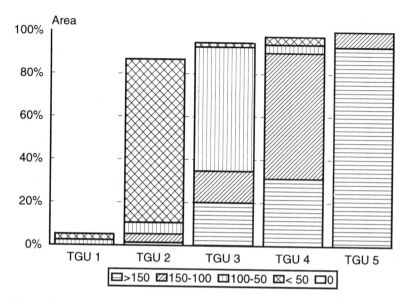

Figure 6.3 Nitrogen input in types of grassland (TGU)

interpreted as evidence of abandonment but it is not. In fact these grasslands are under-utilized from a field-scale agronomic point of view and, with more care, the yield would be higher. The farmers managing these grasslands either think that the physical constraints (e.g. slope) are too high to intensify the management regime, or they only need this piece of land at a low level of productivity.

One methodological problem is that for extensively used grassland many different practices may exist and they may change from year to year. Dates of first grazing, winter grazing and stocking rate are more variable than under intensive management that aims at a certain level of production. Intensive grassland is managed in a way similar to maize or other crops. Extensive management aims at gathering what grows or simply to maintain the land-scape 'clean'.

Distribution of type of grassland use within farms

A diversity of grassland use exists within each type of farm defined according to their functioning. This diversity depends upon physical constraints, organization of the herd and its management (management in sub-units) and forage system (maize, grass). The level of nitrogen fertilization and of 'global intensification' (type of grassland use) of grassland, increases from specialized meat farms to mixed dairy–meat farms and then to specialized dairy farms (Figure 6.4).

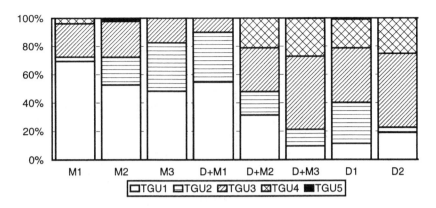

Figure 6.4 Types of grassland use (TGU) in different types of farms classified according to their functioning (M = Meat, D = Dairy, D + M = Dairy + Meat)

Most of the time there are special grazed pastures and both grazed and hay meadows. Farmers make allotments within the herd according to the nature of the production (dairy cows versus beef cattle for example) and the 'production stage': fattening cattle in meat farms, or dairy cows in dairy farms (directly contributing to the main production in the farm) versus steers or heifers, for example.

Sets of grasslands are attributed to sets of animals. The grassland which is most intensively managed (primary function) is used by the set of animals directly contributing to the main production. It is localized on parcels which have no constraints (e.g. dairy cows near farm building, on flat or gently sloping parcels). Grassland which is less intensively managed (secondary function) is assigned to animals of lesser importance: young animals, some beef cattle (within dairy farms). Parcels with many constraints are attributed to these animal sets.

Within meat farms and some dairy–meat farms, the global stocking rate is low and most of the extensive grassland is found on these farms. But the grassland is managed intentionally and the share of both main and secondary grassland is in accordance with the corresponding grazing animals and the local stocking rate. Therefore, a true typology of grassland use would require a precise follow up of the farming operations over several years.

Within dairy farms the global stocking rate is higher than in meat farms, and both extensive and intensive grassland are found. Non-arable land and – more generally – land with many constraints is not abandoned but developed with less costly means (extensive grassland = grassland of secondary function): less inputs, lower stocking rate, grazing animals that need less care. On the other hand, land with less constraints is intensified: ploughed parcels or intensive permanent grassland with a high stocking rate.

113

Diversity of grassland flora as related to management practice diversity

Analysis of the grassland relevés yielded two main gradients in terms of species composition. The first is linked to nutrient richness and the second to humidity. The plot of the first floristic gradient against the first gradient of farming practices, used for the definition of the type of grassland, shows a relationship (Figure 6.5). Within intensive practices (fertilization, silage, etc.) the floristic gradient is very narrow. It expands under extensive practices because differences in the physical environment become the main factor controlling vegetation. In this part of the agricultural gradient the number of species is highest (around 50 and up to 100 or more in chalk grassland).

It is of interest to note that some grassland receiving no fertiliser has a species composition similar to that receiving 100 kg of nitrogen per annum or more. This is possible in this type of landscape because nutrients flow downhill from calcareous nutrient-rich soils into clay-rich soils with high

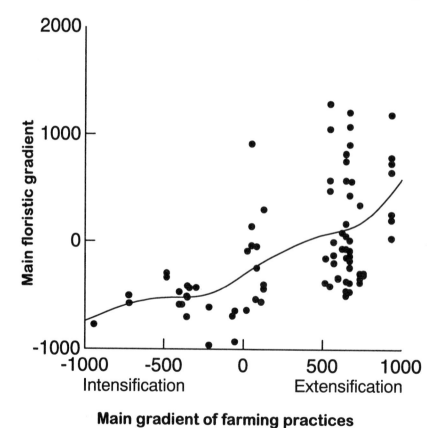

Main gradient of farming practices

Figure 6.5 Relationship between the main floristic gradient in grassland and the main gradient of farming practices

organic matter content. That is to say that even under low input practices the grassland mosaic is very diverse.

Bramble (*Rubus*) patches are present in some grassland. They result from farmers neglect but their origin is unknown. When 'clean' grassland is abandoned for a few years, it is not necessarily invaded by *Rubus*. It seems that the presence of cattle (to open the herb layer) and late grazing (to allow for early growth of *Rubus*) facilitate invasion. Farmers often mow the brambles every few years with a tractor, otherwise it leads to abandonment. The ecological dynamics of those patches have been intensively studied and the reader can find results elsewhere (Baudry, 1991; Burel and Baudry, 1995; Baudry *et al.*, 1997). The main result is that colonization depends on proximal sources of species, as demonstrated for carabids and plants. The pattern is different for spiders.

USE OF STATISTICS AT THE REGIONAL LEVEL

Coarse grain statistics are used at the Département and Région levels in two manners: to test the possibility to extrapolate local results to a broader scale, and to analyze long-term changes in land use. This provides further examples for the need to use scale as a tool of analysis.

Land-use pattern and types of household income

We have shown above the relationship between the farm land use system and the types of household income. Because structural farm and household types are based on data available in agricultural censuses it has been possible to characterize the whole study area, as well as different agricultural regions of Lower Normandy according to the same criteria. Therefore, for example, the effect of household income structure on land-use patterns has been tested in different conditions. In any case, multiple job households, where farming requires at least one full-time worker, have more ploughed land than others (Figure 6.6).

The differentiation among households also explains why there is no massive abandonment of land (Laurent, 1991). The crisis in agriculture is only a part of the economic crisis that has existed in Europe since the mid-1970s. The increase in unemployment rate is a major indicator of the crisis. The huge decrease of farm population after World War Two was partly due to increased possibilities to find a job in industry. Now that this has ended, farmers stay on the land because they have no choice or because they can have a part-time job on the farm that complements income earned elsewhere.

Long-term change in land use

Land-use change was the starting point and casual observations during the research period showed that both abandonment and intensification were taking place. We also noted cases of abandonment for a few years followed by

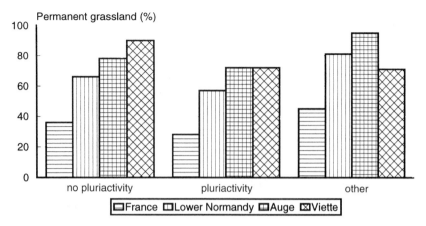

Figure 6.6 Importance of permanent grassland in farms, defined according to the presence of pluriactivity at various scales in France

re-cultivation. To obtain a picture of past changes and of their intensity, we used statistics on agriculture over the last hundred years. A detailed analysis (Baudry, 1992) shows that changes were important and rapid. Normandy, renowned as a grassland country today, was largely used for crops during the last century. Permanent grassland reached its peak in the 1970s, before the effects of the CAP and associated processes of intensification reversed the trend and led to ploughing of grassland (Figure 6.7).

A striking result of this change is the scale dependence of the rates of changes. The longer the period over which the rates of changes are calculated, the less the apparent changes are (see Reenberg and Baudry, Chapter 2). This means that over a short period, changes can be in the opposite direction to the main trend (e.g. cropland reverted to grassland). These changes can be perceived on a year-to-year basis but compensate each other over a few years.

CONCLUSION

Beyond the specific results of this case study, the research yields several general results.

Land abandonment was not found to be a real threat but there is a discrepancy between signs in the landscapes and functioning of farming systems. The situation may stay as such as long as the labour market does not improve. Research in other regions (Laurent *et al.*, 1994) shows that the traditional model of farm production, the son taking up his father's farm, no longer holds. Several trajectories in one's life can explain why people stay or go back to farming in the current economic crisis. Predicting land-use changes with only variables describing the agricultural economic sector is a fallacy. Changes are driven by the economy as a whole.

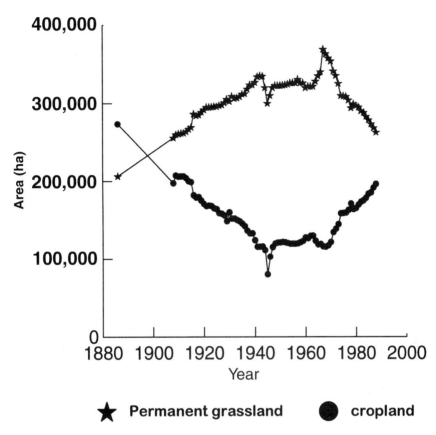

Figure 6.7 Change in permanent grassland and crop area in the Départment of Calvados (Source: Ministry of Agriculture, annual census)

In the Pays d'Auge, permanent grassland was the almost exclusive type of land cover in the 1960s and 1970s. The intensification of agriculture induced by European policies then led to increased ploughing. The trend is uneven among farms; our study demonstrates that the family (farmer's age) and social factors (pluriactivity, professional contacts) play a major role. This means that there are individual and group reactions to policies and – rather than the initial state of land cover – the policy context is much more predictive.

In cattle raising farms the possibility to diversify utilization of land within a farm is important. This permits incorporation of land with high physical constraints within the system and so avoid within-farm land abandonment. Therefore, the maintenance of the visual aspect of the landscape and of grassland-rich flora is ensured. In the Pays d'Auge those species rich grasslands are on the upper slopes of the valleys. We must point out that agri-environmental criteria such as 'average stocking rate' can be very misleading in terms of the impact of cattle on the land, especially for flora. Behind the average there can be a great variation between grassland lightly grazed and grassland fertilised for the grazing of dairy cows and production of silage.

Another important result is the building of a theoretical framework to study land-use changes: the possibility to integrate various disciplines through a spatial and hierarchical approach. The already known complexity of land-use changes can be somewhat elucidated in this manner. It is important to keep in mind that the integration of disciplines was made possible because our theoretical background refers to similar objectives: fields, farms, spatial patterns (landscape or the territory of municipalities) arranged in a hierarchical manner. This enabled us to describe most of them from two points of view: a field as part of a farming system and as part of an ecological system, or a farm as a means of production and as a social structure (household). Such research gives us the means to explore ecological changes at the regional level. Our study exemplifies the coupling of ecological, agronomical and socio-economic approaches to elucidate mechanisms of land-cover/use dynamics as advocated by Turner and Meyer (1994).

ACKNOWLEDGEMENTS

The study was made possible through the continuous financial support of INRA and the Ministry of Environment (Comité Ecologie et Gestion du Patrimoine Naturel). We also thank Dominique Volland and Wanda Sujeska for help during field work.

REFERENCES

Allen, T. H. and Hoekstra, T. W. (1992). *Toward a Unified Ecology*. Columbia University Press, New York

Baudry, J. (1991). Ecological consequences of grazing extensification and land abandonment: role of interactions between environment, society and techniques. *Options Méditerranéennes*, A15: 13–19

Baudry, J. (1992). Dépendance d'échelle d'espace et de temps dans la perception des changements d'utilisation des terres. In Auger, P., Baudry, J. and Fournier, F. (eds) *Hiérarchies et échelles en écologie*. Naturalia Publications, Turriers pp. 101–13

Baudry, J., Alard, D., Thenail, C., Poudevigne, I., Leconte, D., Bourcier, J.-F. and Girard, C.M. (1997). Gestion de la biodiversité dans les prairies dans une région d'élevage bovin:le Pays d'Auge, France. *Acta Botanica Gallica*, **143**: 367–81

Baudry, J., Thenail, C., Le Coeur, D., Burel, F. and Alard, D. (1994). Landscape ecology and grassland conservation. In Haggar, R. J. and Peel, S. (eds) *Grassland Management and Nature Conservation*. British Grassland Society, Aberystwyth pp. 157–66

Boyer, R. (1986). *La théorie de la régulation: une analyse critique*. La Découverte, Paris

Brossier, J., de Bonneval, L. and Landais, E. (eds) (1993). *System studies in agriculture and rural development*. INRA Editions, Paris

Burel, F. and Baudry, J. (1995). Species biodiversity in changing agricultural land-scapes: A case study in the Pays d'Auge, France. *Agriculture Ecosystems and Environment,* **55**: 193–200

Laurent, C. (1991). Place de l'activité agricole dans l'espace, l'exemple d'une région agricole de Normandie, le Pays d'Auge. *Economie Rurale,* **202/203**: 34–9

Laurent, C. (1994) L'hétérogénéité des exploitations agricoles à temps partiel. *Cahiers Agricultures,* **3**: 170–4

Laurent, C. and Bowler, I. (eds) (1997). *CAP and the regions: Building a multidisciplinary framework for the analysis of the EU agricultural space.* INRA Editions, Paris

Laurent, C., Langlet, A., Chevallier, C., Jullian, P., Maigrot, J. L. and Ponchelet, D. (1994). Ménages, activités agricoles et utilisation du territoire: du local au global à travers les RGA. Cahiers d'études et de recherches francophones. *Agricultures,* **3**: 93–107

Moati, P. (1989). Les facteurs démographiques, économiques et structurels de l'évolution. *Comptes rendus de l'Académie d'Agriculture de France,* **75**: 9–16

Schoute, J. F. T., Finke, P. A., Veeneklaas, F. R. and Wolfert, H. P. (eds) (1995). *Scenario studies for the rural environment.* Kluwer Academic Publishers, Dordrecht

Thenail, C. (1992). *Fonctionnement des exploitations agricoles du pays d'Auge et utilisation des prairies permanentes.* INA-PG, INRA-SAD, Normandie

Thenail, C. and Baudry, J. (1996). Consequences on landscape pattern of within farm mechanisms of land-use changes (example in western France). In Jongman, R. H. G. (ed.) *Land-use changes in Europe and its ecological consequences.* European Center for Nature Conservation, Tilburg pp. 242–58

Turner II, B. L. and Meyer, W. B. (1994). Global land-use and land-cover change: an overview. In Meyer, W. B. and Turner II, B. L. (eds) Changes in land use and land cover: a global perspective. Cambridge University Press, Cambridge pp. 3–10

Zonneveld, I. S. and Forman, R. T. T. (1989). *Changing Landscapes: an ecological perspective.* Springer Verlag, New York

AGRICULTURAL LAND-USE AND LANDSCAPE CHANGE UNDER THE POST-PRODUCTIVIST TRANSITION – EXAMPLES FROM THE UNITED KINGDOM

I.R. Bowler and B.W. Ilbery

INTRODUCTION

This paper is concerned with one of the main 'driving forces' in agricultural land-use and landscape change, namely state intervention. Central to the argument is the concept that, over the last decade, agriculture in the European Union (EU) has moved into a period of development termed 'the post-productivist transition' (Lowe *et al.*, 1993). Following forty years of expansion in production after the Second World War, the farm sector is being redirected by state (i.e. EU) intervention to reduce its food production, operate with less protection from the state in a competitive international market, provide society with 'environmental goods', and redesign its methods of production to create a more 'sustainable' agriculture for the longer term. A number of impacts on agriculture and its land uses, and hence on rural landscapes, flow from the post-productivist transition. Here the transition is conceptualised as comprising three bi-polar dimensions in agricultural landscapes: intensification–extensification, concentration–dispersion and specialisation–diversification. These dimensions are examined in the context of recent changes in the Common Agricultural Policy (CAP), with empirical evidence of their impact on land use and landscapes presented at three spatial scales: regions within the EU, counties within England, and a sample of farms from the county of Lincolnshire in the east of England.

AGRICULTURAL LAND-USE AND LANDSCAPE CHANGE UNDER PRODUCTIVISM

Land use, and hence landscape change within productivist agriculture (i.e. maximising production) in the EU has been subjected to extensive study in a number of disciplines and reviewed by Bowler (1987). While there are complex national and regional variations, agricultural land-use trends associated with the productivist period of development can be simplified under three dimensions: intensification, specialisation and concentration (Bowler, 1985). Each of these dimensions can be examined at a variety of spatial scales: for example, through agricultural census units such as individual countries, administrative regions, local administrative districts, and individual farm businesses. *Intensification* can be measured either by increased farm inputs (e.g. capital and fertiliser) or farm outputs (e.g.

production of cereals or meat) per hectare of agricultural land. All of the countries in the EU, for example, were characterised in the four decades leading to the mid-1980s by rapid and sustained increases of both inputs to and outputs from agricultural land. Increased *specialisation*, on the other hand, can be measured as the percentage of the total value of agricultural output contributed by the various farm products in a census unit or, as a surrogate measure, by the percentage of all agricultural land occupied by each crop, including grass. Regions and farms have tended to increase their specialisation in the products and land uses for which they have a comparative economic advantage. The trend towards increased *concentration* in agricultural land use describes the increasing proportion of total production of a crop or livestock from a decreasing number of census units. Concentration within various *land uses* is underpinned by the concentration of *land* at the farm level through the process of scale enlargement.

The logic of the productivist period in the member states of the EU owed much to the interaction between market capitalism and state intervention through agricultural policy: initially that interaction took place within individual nation states, but more recently the context has been the EU (formerly European Economic Community), with countries acting together under the CAP. Under both policy-making contexts, the expansion of food production was accorded a priority as a strategic objective, financial support was provided for the farm sector, and protection was given against cheaper food imports from competitor countries. A dense web of regulations was created to support agricultural productivism (Bowler, 1985), including measures such as internal price support for farm products (target prices, intervention prices, monetary compensation amounts), export subsidies on farm products, variable import levy protection against imports, direct income supplements for farmers in Less Favoured Areas (LFA), investment grants for farm modernisation, farmer training and farmer retirement. In expanding food production from their businesses, farmers responded positively and rationally to enhanced product prices, guaranteed markets, and the continuous stream of new farm technologies emerging from the industrial sector, including agri-chemicals, farm machinery and food manufacturing. Thus the term 'industrialised farming' has been used to summarise the characteristics of productivist agriculture and its associated land uses and landscapes (Troughton, 1986).

Uneven regional trajectories of agricultural land-use and landscape change, and increasing differentiation between rural areas, resulted from the operation of these three processes under agricultural productivism (Jansen and Hetsen, 1991; Marsden *et al.*, 1987). The consequentially rising costs of farm support programmes, unequal benefits to the farm population, and damaging environmental impacts (e.g. water pollution, habitat loss), with their expression in the landscape, are well researched and understood in the UK (for example: Green, 1986; Potter, 1986).

CONCEPTUALISING THE POST-PRODUCTIVIST TRANSITION AND THE ROLE OF AGRICULTURAL POLICY

The exact nature of post-productivist agriculture has yet to be defined by governments and society in the EU but, following Bowler and Ilbery (1996), it is possible to conceptualise the transition to post-productivism as three bipolar dimensions of change:

(1) From *intensification* to *extensification*. Many policy measures introduced under the CAP since the mid-1980s have encouraged once intensive farm businesses to decrease their level of purchased non-farm inputs and become increasingly more extensive in their production. There are important implications for the reduction of inputs in levels of environmental pollution and the restoration of natural habitats in the landscape, especially in the grass-based livestock and cereal sectors.

(2) From *concentration* to *dispersion*. The trend towards increasing polarisation in agriculture, whereby most output becomes confined to fewer and larger farm businesses and regions, may be reversed by the 1992 revisions to the CAP (Robinson and Ilbery, 1993). Farmers may be encouraged to sub-divide their farm businesses into smaller units, thus dispersing agricultural production. However, there is little current evidence of this and dispersion is the least likely dimension of change to occur.

(3) From *specialisation* to *diversification*. Encouraged by the 'cost-price squeeze' on traditional farm products, together with revised policy measures under the CAP, farm businesses are seeking to develop new sources of income through different types of agricultural and non-agricultural diversification. This has necessitated a move away from farming systems where a large proportion of total output is accounted for by a particular product and a 'monoculture'. Such diversification trends enable more diversified land-use systems to be created, with associated implications for a more varied landscape.

These dimensions in the post-productivist transition of EU agriculture must be placed in the context of three political developments: first, reforms of the CAP since 1992; second, the General Agreement on Tariffs and Trade (GATT) negotiations which were finalised in 1993; and third, the increasing convergence between agricultural and environmental policy in the EU. Together they form the context of state regulation for the post-productivist transition.

Reform of the CAP can be traced back to 1984 and a series of measures designed to control agricultural production in the EU. These included guaranteed thresholds for cereals, sunflower seeds, processed fruit and raisins, and compulsory quotas in the milk sector (Ilbery, 1992a). A year later the European Commission's Green Paper, *Perspectives for the CAP*, initiated a wide-ranging debate on the future of European farming and suggested guidelines for reform of the CAP, with a basic aim of reducing surpluses. Yet it was not until 1988 that the European Commission announced three co-ordinated reform measures based on a restrictive pricing policy, involving a gradual scaling down of support prices for products in surplus, the application of the principle of producer co-responsibility, and limits on intervention guarantees, coupled with a more stringent policy on quality in order to meet market requirements (Robinson and Ilbery, 1993). Progress in implementing the reforms was slow, even though there were increasing external pressures on the EU (through GATT) to cut both the high guaranteed prices paid to their farmers and the volume of subsidised exports. This necessitated a fundamental rethinking of the CAP. The so-called 'MacSharry proposals' of 1991 advocated, for the first time, substantial cuts in support prices for cereals, with smaller cuts for milk, beef and butter; curbs on sheep farming; extensions of the 1980s reforms that had introduced set-aside and early retirement for farmers; and a system of income aid (compensation) for farmers with small and medium-sized farms.

The MacSharry proposals acted as an important catalyst in the continuing discussions over policy reform. Indeed they occupied an important position in the GATT negotiations and showed that the EU was willing to put forward concrete proposals for effecting such controls. On May 21, 1992 agreement was reached by EU farm ministers on reforming the CAP; the basic elements were price reductions and direct compensation to farmers, irrespective of farm size. Guaranteed prices for cereals were cut by 29 percent over three years. To compensate for this reduction, farmers were to receive area-based direct payments (known as Arable Area Payments in the UK), but only if they set aside at least 15 percent of their arable land (for which they were also to receive a payment). In the beef and sheep sector, a system of quotas limiting the number of livestock on which farmers can receive payments was introduced (Wynne, 1994); milk quotas were to remain in place. All of these changes have been controlled subsequently through the Integrated Administration and Control System (IACS), set up by the European Commission (Regulation 3887/92), whereby farmers have to record all details of crops, livestock and set-aside on an annual basis.

The main impacts of the 1992 reforms have been to replace the price guarantee system by direct income aid payments and to ensure that farm incomes are not tied to (i.e. they are decoupled from) the amount of food produced. Completion of the GATT negotiations in 1993 served to strengthen the CAP reforms, with agreements to reduce price support, export subsidies and border controls applied to agricultural commodities around the world (Arden-Clarke, 1992). Indeed, the EU agreed to cuts of 21 percent and 35 percent in the volume and value of subsidised exports, respectively, and to limit the area of oilseeds to 5.13 million ha. Free access to the EU market is also to be allowed for imports of 3 percent (and eventually 5 percent) of the internal consumption of agricultural commodities produced within the EU. According to Naylor (1995: 283), the overall effect of the CAP reforms and

GATT agreement is likely to be a 'redistribution of aid to the less intensive and less well off areas, particularly within the livestock sector, and hence a contribution to the reduction in regional income disparities'.

Although the reform measures have been economically-driven, Regulation 2078/92 required individual member states to implement an 'agri-environmental' package ('multi-annual zonal programme') by the end of 1993, with the following aims: to reduce the use of fertilisers and plant protection chemicals in intensive agricultural systems; to maintain 'environmental goods' in the form of wetlands, moorlands, heathlands and woodlands; to manage abandoned farmland; to introduce long-term set-aside for 'surplus' farmland; to encourage the development of more sustainable, lower input–output farming systems such as organic farming and extensive livestock farming; and to manage farmland for public access and leisure activities. Overall, these measures are expected to yield reduced food production, as well as environmental benefits, with consequences for agricultural landscapes in the post-productivist transition.

EVIDENCE OF THE POST-PRODUCTIVIST TRANSITION IN AGRICULTURAL LAND USE AND LANDSCAPES IN THE EU REGIONS

Emerging evidence of the post-productivist transition, expressed in agricultural land use and landscapes, suggests spatial differentiation through the mediation of such features as farm size, farm type, resource endowment, land ownership and land occupation. Figure 7.1 shows the differentiation of agriculture in the EU regions using available agricultural census data for 1979 and 1987. For each region, the intensification–extensification dimension has been measured by change in the standard gross margin (SGM) per hectare of agricultural land; the concentration–dispersion dimension has been measured by change in average farm size (hectares); and the specialisation–diversification dimension is measured by an index of specialisation for five major agricultural land uses (cereals, industrial crops, fruit/vegetables, vines and grassland). The continuing development of the 'industrial model' of agricultural development (Group 1) is evident in most of England, central France, The Netherlands and Denmark, and in certain regions within (west) Germany and Italy. For all the other regions, however, there is evidence of at least one of the extensification/diversification/dispersion trends in operation, but in a highly differentiated pattern between regions. For example, evidence of more extensive agriculture (Group 2) can be seen in northern and southern regions of France and in northern England. However, agricultural census data will need to be available for the mid-1990s before more certain long-term trends can be identified and the causes of spatial differentiation established.

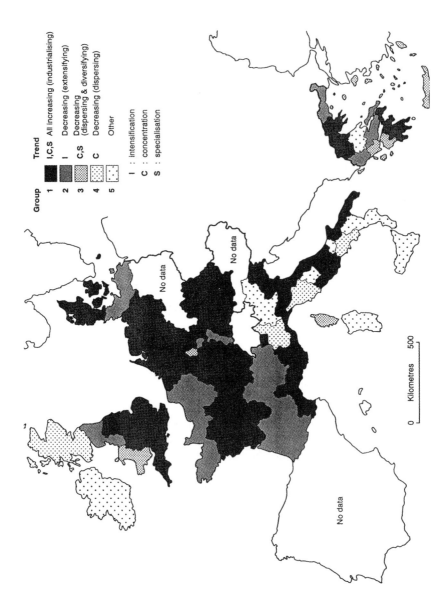

Figure 7.1 A typology of restructuring trends (1979–87)

FROM INTENSIFICATION TO EXTENSIFICATION IN AGRICULTURAL LAND USE AND LANDSCAPE IN THE UK

At least three developments can be identified in the agricultural land use of the UK in support of the concept of extensification under the post-productivist transition: set-aside (land diversion), organic farming, and the production of environmental goods. Each development has its landscape consequences.

Set-aside

As one element in the transition to post-productivism in EU agriculture, Regulation 1094/88 attempted to cut agricultural surpluses by encouraging farmers to take land out of arable production altogether (set-aside). While policies for the extensification of production never really materialised in the UK, the diversion of land out of arable farming was introduced in Britain in 1988. Set-aside was voluntary for individual farmers, although compulsory for member countries of the EU. In return for setting aside 20 percent or more of their arable land (defined as land devoted to the growing of supported crops) for five years, farmers received compensation of between 100 and 600 ECU per ha per year. The retired land could be left fallow or used for either woodland or non-agricultural purposes. Effectively, farmers were being paid not to farm on 20 percent of their arable land, while at the same time they were receiving high guaranteed prices for producing as much as possible on the remaining 80 percent.

Not surprisingly for a voluntary scheme, the amount of land set aside was low: between 1988 and 1992 just 4,478 farmers agreed to set aside 154,579 ha, mainly as permanent fallow. This represented less than 5 percent of the area devoted to cereal farming. A major problem with the Scheme was that it allowed farmers to set aside their poorest arable land and intensify production on the remaining arable land. Consequently, production tended not to fall in proportion to the amount of land retired (a phenomenon known as slippage in the USA). The result was a disappointing and spatially uneven pattern of uptake (Figure 7.2a). Rather than being concentrated in the 'core' cereal producing areas of East Anglia and Humberside, set-aside favoured those areas with a weak competitive ability in cereals (i.e. marginal cereal areas) and a strong competitive ability in both mixed and part-time farming (Bowler and Ilbery, 1989). Thus counties in close proximity to London showed the highest rates of set-aside, possibly reflecting market opportunities for off-farm employment (other gainful activities – OGAs) and diversification activities (Ilbery, 1990).

Another major weakness of the voluntary set-aside scheme was its lack of consideration for the environment/landscape. Few environmental benefits can accrue from a scheme that takes very small and highly fragmented parcels of land out of production; far greater benefits would have occurred by targeting large tracts of land suitable for conservation. Following the reforms of the

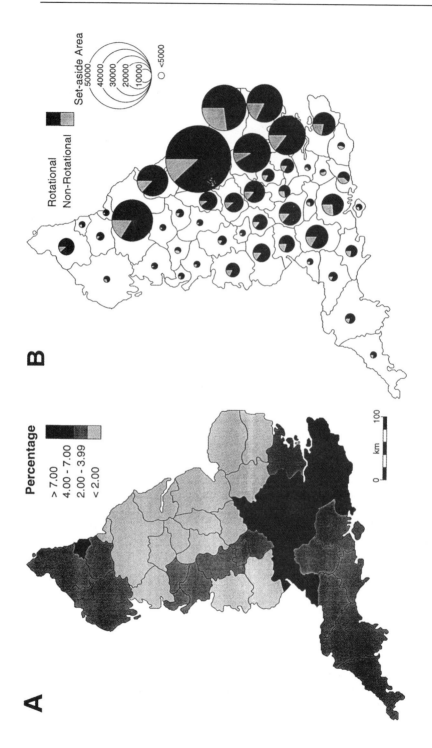

Figure 7.2 Distribution of set-aside in England: (a) area set-aside as a percentage of supported arable crops under the voluntary scheme, 1992; (b) area of rotational and non-rotational set-aside under the compulsory scheme, 1994

CAP in 1992, set-aside effectively became compulsory for all farmers (except for those producing a very small output of cereals). In return for setting aside arable land, farmers became eligible to claim income aid through the Arable Area Payments Scheme (AAPs). Two options became available to farmers in Britain: rotational and non-rotational set-aside. The former required farmers to set-aside 15 percent of their arable area on a rotational basis for each of six years, while the latter involved setting aside 18 percent of the total area on which AAPs are claimed for five years. In 1994, 553,175 ha were set aside, of which just 17.5 percent was through the non-rotational scheme. Geographically, the 'core' cereal areas are now setting aside more arable land (Figure 7.2b), but this is mainly on a rotational basis which yields few environmental/ landscape benefits.

Set-aside measures continue to be criticised for their lack of environmental concern (Craighill and Goldsmith, 1994). However, when linked to the 1993 agri-environmental package in the UK (see below), the 20-year Habitat Improvement Scheme presents farmers with an opportunity for the construc- tive use of set-aside land. Nevertheless, it is likely that less than 10,000 ha will be attracted into this voluntary scheme and until environmental conditions are attached to livestock headage payments and AAPs, through a system of cross- compliance, set-aside will remain a limited tool of environmental management.

Organic farming

Turning to the development of low input–output agriculture, organic farming throughout the UK now receives financial assistance: farmers converting their land use from conventional to organic farming practices now receive financial compensation under the 'Organic Aid' Scheme for their reduced income during the transition period. But there has been a muted response in intensive crop farming areas. Rather organic farming continues to develop in three main areas (Bowler and Ilbery, 1993): adjacent to the main concentration of organic food consumers in the London conurbation; on small farms in western parts of Wales; and in the Vale of Evesham where horticulture has been a traditional farming practice. To date, there has been insufficient demand in the UK from con- sumers to support a marked expansion in the production of organic farm produce and much of the organic food that is consumed is imported into the country.

Environmental goods

The third example from the UK of extensification under the post-productivist transition is provided by payments to farmers for the purpose of environ- mental protection and countryside management. The common thread linking these agri-environmental schemes is that they are mainly voluntary, rely on incentive payments, and are limited to a restricted number of designated areas. As Potter (1993) demonstrates, the Environmentally Sensitive Areas (ESA) programme covers only 15 percent of the UK. Within each of the 22 ESA, farmers receive financial compensation for carrying out farming practices which benefit nature conservation, the landscape and the conservation of historic features. Financial compensation is provided for reduced livestock densities (beef cattle and sheep), reduced applications of nitrogenous fertilisers, the management of woodlands and herb-rich meadows for nature conservation, and the repair and maintenance of field walls and traditional farm buildings. In addition there is an expanded Nitrate Sensitive Areas (NSA) scheme which offers remedial environmental measures to combat groundwater pollution. Farmers in the 32 Nitrate Sensitive Areas (NSA) – now designated as Nitrate Vulnerable Zones (NVZ) under the EU Nitrate Directive 91/676/EEC – receive financial compensation to limit their fertiliser applications, convert arable land to grassland, or control nitrate leaching through crop rotations. Here the objective is to reduce the pollution of surface water and groundwater resources from nitrates.

The landscape impacts of these agri-environmental measures in the UK awaits detailed research, but the measures are unlikely to be fully effective while remaining 'bolted-on' to an economically-driven policy (Robinson, 1991; Ilbery, 1992b). Indeed, Potter (1993) emphasises that just 2 percent of the farm budget is likely to be spent on the agri-environmental schemes, with the remaining 98 percent being reserved for supporting prices and funding compensation measures like set-aside. Moreover, outside the designated areas, variations to the productivist model of farming depend on decision making by individual farmers. For example, farmers can vary their decisions on whether to place set-aside land into short-term (rotational) or long-term (conservation) uses, including a range of non-food crops such as biofuels. Similarly, farmers have voluntary entry into other state-financed schemes to promote the provision of 'environmental goods', such as the 'Habitat', 'Moorland', 'Farm Woodland Premium', 'Countryside Stewardship' and 'Countryside Access' Schemes (Ilbery, 1992a) The pattern of response varies by individual farm and between farming regions.

FROM SPECIALISATION TO DIVERSIFICATION IN AGRICULTURAL LAND USE AND LANDSCAPE IN THE UK

Farm diversification refers to the development of non-traditional farm enterprises and involves a diversion of resources (land, labour and capital) which were previously committed to conventional farming activities. It needs to be distinguished from the wider concept of pluriactivity, which involves the generation of income from both on-farm diversification and off-farm OGA (MacKinnon *et al.*, 1991). Ilbery (1991) identified two types of farm diversification: agricultural and structural. To date, *agricultural diversification* has been concerned mainly with a search for alternative crops, such as linseed and evening primrose; but relatively small markets have been found for these products. Rather, forestry has been identified as the alternative land use with the greatest potential for absorbing large areas of farmland. Commercial timber production, biofuel production, nature conservation, recreational use and aesthetic contribution to the landscape have all been identified as positive outcomes from afforestation.

Nevertheless, more research attention in the UK has been given to *structural diversification*, although the environment/landscape impacts of such developments as pony-trekking and golf courses have yet to be researched. While the UK has a long history of farm diversification, it has developed quite rapidly since the mid-1980s and especially after the introduction of a Farm Diversification Grant Scheme (FDGS) in 1988. The FDGS offered financial assistance towards the development of different types of structural diversification. As with many grant schemes offered to farmers, uptake of the FDGS was slow and spatially uneven (Ilbery and Bowler, 1993). The highest concentrations of uptake have occurred in the main tourist areas in southwest and upland Britain (for farm-based accommodation) and around urban fringe areas in central and southern England (for farm-based recreation).

EVIDENCE OF THE POST-PRODUCTIVIST TRANSITION AT THE FARM LEVEL

The previous discussion has focused on the role of state (i.e. EU) intervention in promoting the three bi-polar dimensions of the post-productivist transition. A recurring theme has been the role of individual farmer behaviour in responding to instruments of state intervention, with resulting impacts on agricultural landscapes and environments (Potter, 1986). This theme is now explored in greater depth by taking a random sample of farmers from one lowland county of England – Lincolnshire – and examining their recent farm business decisions as regards the interaction between agriculture and the environment. Three study areas, with 25 farmers from each area, have been used in the analysis to evaluate the contention that the greater differentiation of agriculture is based in part on the geographical context of farmer decision making. Here 'geographical context' is interpreted in terms of physical environment or 'nature' (Figure 7.3). Each study area comprises a different geographical context: the Fens, the Clays and the Wolds of Lincolnshire. The Fens study area occupies low-lying, flat land with poor natural drainage and gley soils derived from reclaimed marine alluvium (Wallasea 2 Association); the Clays study area lies in the valley of the River Witham and comprises flat land with clay soils developed from chalky tills (Bales 1 Association); the Wolds study area is located on a rounded, dissected, north–south chalk escarpment which rises to 200 m, with well-drained, calcareous, loam soils (the Swaffham Prior Association on the upper slopes and the Cuckney 2 Association on the lower slopes).

Reflecting variations in the physical environment, the three study areas are differentiated by farm size and type. The Wolds study area, for example, has

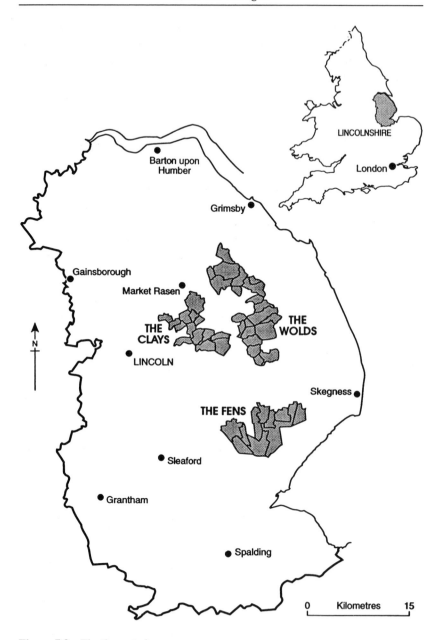

Figure 7.3 The three study areas

a higher proportion of large farms (i.e. over 300 ha) compared with the other two study areas whose farm-size structure is similar and characterised by farms of less than 150 ha. The Wolds area also has a higher proportion of its farms specialised in the production of cereals (wheat or barley) compared with the other two areas. The Fens study area features farms producing sugar beet, vining peas and brassicas, while the Clays area has a higher proportion of farms with livestock, oilseed rape and potatoes.

Farm-level data on indicators of the post-productivist transition confirm the differentiation of agriculture both within and between the study areas. For example, data were collected in 1993 on farm trends over the previous 10 years as regards the three bi-polar dimensions discussed earlier. The results are shown in Table 7.1. Looking first at the intensification–extensification dimension, a majority of the respondents reported that their inputs of fertilisers had fallen over the previous 10 years, an extensification trend most developed in the Clays and least developed in the Wolds study area. However, significant percentages of farmers claimed that their fertiliser inputs had increased, especially in the Fens study area. For pesticides, the main response was a continued increase in their use, especially on the vegetables of the Fens; only a minority of farmers reported a reduction in their use of agrichemicals.

Turning to the concentration–dispersion dimension, most farmers reported either no change in the area of their farms or a continued increase in farm size, again especially in the Fens. The dispersion dimension (towards a smaller farm size) remains a minor feature of agriculture in all the study areas. The specialisation–diversification dimension is equally developed in the three study areas, with under 30 percent of farm businesses having diversified over the last 10 years.

The post-productivist transition involves the creation of 'environmental goods', set within the three bipolar dimensions of change. Again, in examining the incidence of certain environmentally beneficial changes on the sampled farms over the last 10 years (Table 7.2), a similar pattern of variability both

Table 7.1 Indicators of the post-productivist transition in the farm sample (percent of farms in each study area)

Dimension	Indicator	Trend	Fens	Clays	Wolds
	Fertilisers	increase	36	28	32
I–E		decrease	44	48	32
		no change	20	24	36
	Pesticides	increase	56	40	44
I–E		decrease	16	24	12
		no change	28	36	44
		increase	64	32	36
C–D		decrease	12	12	8
		no change	24	56	56
S–D	Diversification increase		20	28	28

I–E: Intensification–Extensification; C–D: Concentration–Dispersion; S–D: Specialisation–Diversification

133

within and between the study areas emerges. Overall, the production of environmental goods outweighs their destruction, suggesting the emergence of more sensitive environmental behaviour amongst many but not all farmers. However, the varied pattern of farmer behaviour confirms one of Potter's (1986) conclusions about the role of 'investment styles' in countryside change: namely, farmers tend to focus on single projects of landscape enhancement and the environmental impact is produced by 'incremental' rather than 'systematic' investment planning. Negative impacts on environmental goods have a lower level of incidence in all the study areas, except for continuing increases in field sizes and the incidence of in-filled ponds in the Clays study area, and the erection of farm buildings in all the study areas.

Table 7.3 investigates other facets of variation in farmer behaviour within and between the three study areas by focusing on farmers' (voluntary)

Table 7.2 Indicators of the production and destruction of environmental goods on the sample farms (percent farms with each indicator)

Indicator	Fens	Clays	Wolds
Production			
Increased access by the public	24	36	56
Increased bird population	56	60	32
Made a new pond	20	28	32
Built new hedgerow	36	28	32
Planted new woodland	52	56	20
Destruction			
Filled-in a pond	4	20	0
Increased field size	12	28	16
Removed a hedgerow	8	16	16
Erected a new farm building	56	80	52

Table 7.3 Farmer commitment to environmental conservation (percent farms with each indicator)

Indicator	Fens	Clays	Wolds
Has undertaken conservation work with grant aid	54	84	72
Has undertaken conservation work without grant aid	36	36	36
Has been on a training course for conservation	28	32	52
Has a future conservation plan for trees	32	48	56
Has a future conservation plan for hedgerows	24	40	28
Carries out conservation work mainly for enjoyment	32	36	40
The meaning of 'conservation' is to:			
create new habitats	56	44	68
retain existing habitats	84	80	96
limit environmental damage	40	44	52

response to state schemes of financial assistance to promote the production of 'environmental goods' (e.g. the Farm and Conservation Grant Scheme and the Farm Woodland Grant Scheme). A high proportion of farmers in the Wolds study area claimed to have carried out conservation work with grant aid but, as described already, their investment has been limited to narrowly defined projects of environmental conservation. Nevertheless, in all study areas over half the farms had implemented some form of conservation work, although rather less without grant aid. Wolds farmers had also attended training courses on conservation in larger numbers, had a higher incidence of future conservation plans for trees, but not hedgerows, and were more highly motivated by enjoyment in carrying out their conservation work, as compared with the functional role of conservation in the farm business, or a sense of idealism or responsibility towards the environment. Perhaps because of their training courses, Wolds farmers more often reported the meaning of 'conservation' in terms of both retaining and creating new habitats compared with farmers in the other two study areas.

For all farmers, however, 'environmental conservation' is interpreted, or 'negotiated' (Ward and Munton, 1992: 133), mainly in terms of habitat creation and retention, rather than as a means of reducing the wider environmental damage created by productivist farming systems. This negotiated interpretation reflects the way in which the state employs 'positive', voluntary schemes of grant aid to promote habitat conservation (incentive-based policies), but uses 'negative', mandatory regulations to address broader problems of environmental conservation (e.g. water pollution from fertilisers and pesticides). Through this dual policy approach, state intervention plays an important role in landscape production by the farm community, although Bishop and Phillips (1993) provide a discussion of the wider range of incentive policies that could be employed.

A final insight into agriculture in the post-productivist transition can be gained by examining farmers' beliefs about possible future trends. Table 7.4 summarises this approach and shows three areas of agreement within the farmer sample (e.g. less price support, more farm extensification, and more restrictions on the use of fertilisers) but a number of areas of disagreement (e.g. farm diversification, organic farming, conservation grants). Nor are the differences systematically distributed between the three study areas. For example, farmers in the Fens study area are sensitive to the further development of nitrate–sensitive areas; farmers in the Clay study area, by comparison, foresee a limited future for permanent set-aside but greater potential in conservation grants. Overall, however, farmers see a future with greater restrictions on their farming practices but more emphasis on environmental measures. This interpretation of the future will remain a source of tension and conflict within the farm community. The remaining data in Table 7.4 clearly show that a majority of farmers are still not persuaded of the damage that their farming practices inflict on the environment, despite unanimously seeing

Table 7.4 Farmer beliefs in the future of agriculture under the post-productivist transition (percent farmers agreeing with each proposition)

Proposition	Fens	Clays	Wolds
In the future there will be increased:			
price support for arable crops	0	24	4
farm diversification	52	40	64
farm extensification	48	52	56
nitrate-sensitive areas	92	64	68
organic farming	35	22	19
restrictions on fertiliser use	64	64	44
permanent set-aside	72	48	76
conservation grants	76	80	44
Farmers are stewards of the countryside	100	100	100
Agriculture is the most important use of the countryside	84	96	96
Landscape features should have more protection	44	52	52
Farm technology damages the countryside	32	56	40
Fertilisers damage the environment	20	12	24

themselves as stewards of the countryside and recognising agriculture's dominant role in rural land use. This conflict between society's new environmental goals and agriculture's continuing productivist goals remains unresolved and will limit the development of the three bi-polar dimensions in the post-productivist transition. Indeed a further 'de-coupling' of the CAP from productivist agriculture is probably necessary to alter farmer behaviour more fundamentally as regards the environmental sensitivity of their agricultural practices.

CONCLUSION

The broad theme running through this analysis has been the promotion of the post-productivist transition in agriculture by state (i.e. EU) intervention and the uneven impact on agriculture and its associated landscapes. Three principal processes have been identified which, together, are bringing about further differentiation within agriculture: namely, the 'regionalisation' of state intervention in promoting extensification and diversification in designated areas (e.g. ESA and NSA); the greater distinction being drawn between the production of food and environmental goods; and variations in farmer behaviour towards the environment, both within and between farming regions, as individuals follow the 'flexible accumulation' increasingly required by international, competitive market forces.

In the short term, the post-productivist transition is being driven by the over-production of food relative to domestic demand in the EU, and farmland is widely described as being in 'surplus'. Estimates by Lee (1991) for 10 member states of the EU indicate a 'surplus' of farmland varying from 9 to 37 million hectares depending on the assumptions made about population growth in the EU (i.e demand), rates of increase in agricultural productivity (i.e. supply), the level of agricultural inputs per hectare, and the level of financial support under the CAP for subsidised exports. For example, farmland 'surpluses' decrease if lower input–output farming systems are assumed to develop, if subsidies for non-food and biofuel crops (energy) under the CAP increase, or if agricultural production at market prices can be exported from the EU to feed the world's growing population. On the other hand, farmland 'surpluses' increase if yield-increasing developments in plant and animal genetic engineering (the biotechnology revolution) are taken into account.

Faced by farmland 'surpluses', it is now more helpful to consider land use in 'rural' rather than 'agricultural' terms: the main theme to emerge from the published literature on this subject is the diversification of land use and landscape from agriculture in favour of forestry (including short-term coppice for biofuel production), nature conservation, recreation, and non-food crops. New regional land-use patterns and landscapes are developing in the EU and it seems likely that new emphases will develop in different regional farming systems. However, as this paper has shown, much depends upon the 'environmental endowment' of each region and the ways in which that endowment is interpreted by the farmers/land occupiers in the region. In those systems with the most favourable land capability as regards crop production, low-cost, internationally competitive, productivist farming systems will be maintained, but with a greater emphasis on non-food crops, including biofuels for energy production. These regions are likely to retain many of the landscape features of productivist agriculture, including intensive farming practices, specialisation and large-scale farming. Farmers in rural regions near to urban concentrations, however, will be able to exploit their locations by providing services to the urban population, such as recreational activities, direct marketing and living space (residential). In addition, farm households in these locations are likely to be mainly pluriactive. Farmers in hill and mountain regions, on the other hand, may be expected to exploit the quality of their landscapes by providing services for tourism and recreation, together with commercial forestry. Finally, farmers in these and other regions will be able to produce 'environmental goods' through nature conservation and the preservation of valued habitats such as wetlands, moorlands, heathlands and farm woodlands. Within these broad trends, the present analysis has confirmed the considerable farm-to-farm variation in response to the post-productivist transition: features such as farm size, farm family stage, farm business stage, farm type and farming/environmental beliefs all serve as mediators of large-scale or macro driving forces of change.

The convergence of agricultural and environmental policy in the EU underpins these potential developments in land use and landscape, a trend seemingly assured for the medium term by the political agreement reached on the need for 'sustainable development' at the 1993 Rio 'Earth Summit'. Indeed the reregulation of agriculture based on environmental, food health and animal welfare criteria is likely to be strengthened. Nevertheless, a number of problems in the agricultural policy of the EU have still to be addressed, these include: large farms still benefit disproportionately from the CAP; supervision costs of the revised CAP have risen; the new measures have not reduced the EU budget for agriculture; increases in crop and animal yields from the application of new technologies are overtaking savings from set-aside; and the CAP has not yet been sufficiently 'decoupled' from productivist farming. Consequently, further revisions to the CAP seem inevitable when the next policy review takes place; these revisions will have additional, but as yet unknown, consequences for rural land uses in the EU and their associated landscape features.

REFERENCES

Arden-Clarke, C. (1992). Agriculture and environment in the GATT: integration or collision? *Ecos*, **13**: 9–14

Bishop, K. D. and Phillips, A. C. (1993). Seven steps to market: the development of the market-led approach to countryside conservation and recreation. *Journal of Rural Studies*, **9**: 315–38

Bowler, I. R. (1985). Intensification, concentration and specialisation in agriculture – the case of the European Community. *Geography*, **71**: 14–24

Bowler, I. R. (1987). The geography of agriculture under the CAP. *Progress in Human Geography*, **11**: 24–40

Bowler, I. R. and Ilbery, B. W. (1989). The spatial restructuring of agriculture in the English counties, 1976–1985. *Tijdschrift voor Economische en Sociale Geografie*, **80**: 302–11

Bowler, I. R. and Ilbery, B. I. (1993). Sustainable agriculture in the food supply system. In Nellis, M. D. (ed.) *Geographic Perspectives on the Social and Economic Restructuring of Rural Areas*. Kansas University, Kansas, pp. 4–13

Bowler, I. R. and Ilbery, B. W. (1996). The regional consequences for agriculture of changes to the Common Agricultural Policy. In Laurent, C. and Bowler, I. (eds) *CAP and the Regions: Building a Multidisciplinary Framework for the Analysis of the EU Agricultural Space*. INRA, Versailles, pp. 103–18

Craighill, A. and Goldsmith, E. (1994). A future for set-aside? *Ecos*, **15**: 58–62

Green, B. (1986). Agriculture and the environment. *Land Use Policy*, **3**: 193–204

Ilbery, B. W. (1990). Adoption of the arable set-aside scheme in England. *Geography*, **75**: 69–73

Ilbery, B. W. (1991). Farm diversification as an adjustment strategy on the urban fringe of the West Midlands. *Journal of Rural Studies*, **7**: 207–15

Ilbery, B. W. (1992a). From Scott to ALURE – and back again? *Land Use Policy*, **9**: 131–42

Ilbery, B. W. (1992b). Agricultural policy and land diversion in the European Community. In Gilg, A. (ed.) *Progress in Rural Policy and Planning*, Volume 2. Belhaven Press, London, pp. 153–66

Ilbery, B. W. and Bowler, I. R. (1993). The Farm Diversification Grant Scheme: adoption and non-adoption in England and Wales. *Environment and Planning – Government and Policy*, **11**: 161–70

Jansen, A. and Hetsen, H. (1991). Agricultural development and spatial organisation in Europe. *Journal of Rural Studies,* **7**: 143–51

Lee, J. (1991). Land resources, land use and projected land availability for alternative uses in the EC. In Brouwer, F., Thomas, A. and Chadwick, M. (eds) *Land use changes in Europe.* Kluwer Academic, Dordrecht, pp. 1–20

Lowe, P., Murdoch, J., Marsden, T., Munton, R. and Flynn, A. (1993). Regulating the new rural spaces: the uneven development of land. *Journal of Rural Studies*, **9**: 205–22

MacKinnon, N., Bryden, J., Bell, C., Fuller, A. and Spearman, M. (1991). Pluriactivity, structural change and farm household vulnerability in western Europe. *Sociologia Ruralis*, **31**: 58–71

Marsden, T., Whatmore, S. and Munton, R. (1987). Uneven development and the restructuring process in British agriculture: a preliminary exploration. *Journal of Rural Studies*, **3**: 297–308

Naylor, E. (1995). Agricultural policy reforms in France. *Geography*, **80**: 281–3

Potter, C. (1986). Processes of countryside change in lowland England. *Journal of Rural Studies*, **2**: 187–95

Potter, C. (1993). Pieces in a jigsaw: a critique of the new agri-environment measures. *Ecos*, **14**: 52–4

Robinson, G. (1991). EC agricultural policy and the environment: land use implications in the UK. *Land Use Policy*, **8**: 301–11

Robinson, G. and Ilbery, B. W. (1993). Reforming the CAP: beyond MacSharry. In Gilg, A. (ed.) *Progress in Rural Policy and Planning.* Volume 3. Belhaven Press, London pp. 197–207

Troughton, M. J. (1986). Farming systems in the modern world. In Pacione, M. (ed.) *Progress in Agricultural Geography.* Croom Helm, London, pp. 93–123

Ward, N. and Munton, R. (1992). Conceptualising agriculture–environment relations. *Sociologia Ruralis*, **32**: 127–45

Wynne, P. (1994). Agri-Environment Schemes: recent events and forthcoming attractions. *Ecos*, **15**: 48–52

CHAPTER 8

CHANGES OF RURAL LAND USE WITHIN AN AGGLOMERATION – LEIPZIG–HALLE AS AN EXAMPLE

R. Krönert

INTRODUCTION

Land use and land-use changes in agglomerations (large urban regions) show a number of distinctive features: there is generally a very intensive utilisation of arable land; agriculture finds itself exposed to the forces of urbanisation; and environmental pollution caused by industry, traffic and the cities combines with that from agriculture itself. In addition, in the urban region of Leipzig–Halle, there has been severe damage to the landscape caused by the mining of brown coal. Also, after the collapse of the GDR in 1989, and the reunion of Germany in 1990, the Leipzig–Halle agglomeration was exposed to changes in its agricultural production and structure, as reflected in land-use changes (Umweltbericht Sachsen, 1991, 1994).

The Leipzig–Halle agglomeration consists of the core of Leipzig (481,000 inhabitants in 1994), Halle (299,884 inhabitants, 1992) and Dessau (94,528 inhabitants, 1992), as well as the surrounding districts (communities) which have a total of 2,683,362 inhabitants including the core areas. Between 1989 and 1992 the population of the different areas decreased by 6 to 8 percent. The reasons for this loss of population are to be found mainly in emigration to western Germany and a surplus of deaths over births. Today, the net migration rate is nearly in balance but the death rate still exceeds the very low birth rate (less than 6 per 1,000 inhabitants). The number of industrial workers has decreased to a third, and the number of workers in agriculture to less than a quarter of those in employment. About 15 percent of all people able to work are unemployed. Older citizens have retired early and others take part in further vocational training. These groups together form another 15 percent of the economically active population.

The larger part of the Leipzig–Halle agglomeration is situated in the fertile loess region, as reflected in the land use of the districts (Table 8.1). The percentage of land in arable, especially in the region around Leipzig and Halle, is high, whereas the proportion under forests/woodland is very low.

In 1989 the Leipzig–Halle region was in an extremely poor environmental state: the air was overloaded with dust, SO_2 and harmful organic substances; the per capita emission of CO_2 reached a maximum level; rivers and lakes were polluted; and a large number of old, contaminated industrial sites, as well as active waste-disposal sites, contaminated local soils and groundwater. Arable land was over-fertilized with nitrogen and phosphate, while flood

Table 8.1   The percentage of land use by category in the districts of Leipzig–Halle–Dessau

District	Agricultural land (%)	Forests and woodland (%)	Wasteland, mined area (%)	Water bodies (%)	Other economic areas (%)
Leipzig	66.96	15.11	4.67	1.75	11.51
Halle	69.51	13.21	2.23	1.34	11.72
Dessau	57.73	26.99	2.61	2.11	10.56

plains were locally contaminated by heavy metals. The emissions of pollutants was far beyond the regenerative capacity of the landscape and wide areas of landscape were destroyed by extensive open-case mining of brown coal.

More recently, a number of developments have reduced air, water and soil pollution considerably: for example, a decrease in industrial production, closing down of obsolete chemical plants and power stations which burned brown coal, a decrease in open-cast mining of brown coal, a decline in coal stocks, investments in environmental protection (wastewater treatment, flue gas purification), and the partial substitution of brown coal by natural gas and mineral oil. However, new strains on the environment can be observed, such as an increase in motor traffic (NO_x, ozone, smog in summer) and loss of open landscape as a result of construction of large commercial and business areas in the countryside.

Despite environmental pollution, there have been and still are many areas with high landscape diversity, such as the river valleys and flood plains of the Elbe, Mulde, Weiße Elster and Saale rivers with their tributaries, the heath-lands of the Dahlener and Dübener Heide, small lake areas, and landscapes resulting from brown coal open-pit mining. In recent times these areas have been expanded by fallow land from agriculture and together they have allowed many animal and plant populations threatened by extinction to survive. In addition, survival has been aided by the activities of many people working voluntarily for nature protection. As a result, the beaver population of the Elbe has recovered, the white stork is no longer rare and there are nesting sites of the black stork and eagle in the Dübener Heide. The fish populations of the rivers are recovering well and there is a biosphere reserve of European significance in natural flood plain forests on the Elbe.

The chapter now turns to a consideration of recent changes in agriculture in Saxony and their 'driving forces', followed by an analysis of the ecological consequences of land use within the agglomeration of Leipzig–Halle. The effects of brown coal mining and suburbanisation within the rural area of the Leipzig region are then discussed, with particular attention to possible problems of contamination by heavy metals as a result of inputs of brown coal ash over a long time period, and contamination of groundwater by overfertilizing with nitrogen.

CHANGES IN LAND USE AND STRUCTURE OF AGRICULTURE IN SAXONY AND THE 'DRIVING FORCES'

The Leipzig–Halle agglomeration is located partly in the States of Saxony and Saxony–Anhalt and, as a result, comparable statistical data within the agglomeration are not available. But changes in agriculture are similar in all the new federal states, so that data for the State of Saxony can be used representatively. Significant changes occurred in agricultural land use in the former GDR (e.g. Saxony) between 1989 and 1992/1994 (Sächsischer Agrarbericht, 1992, 1994; Krönert, 1994): cultivation of intensively managed crops decreased by about 75 percent; growing of vegetables around cities nearly disappeared; and production of forage crops decreased while grain and oil seeds increased (Table 8.2).

In 1992, fallow land was about 5 percent of arable land, although on fertile soils around Leipzig the proportion was smaller. The percentage of fallow land within the arable land area in Saxony increased to 11.5 percent in 1993 and 12 percent in 1994. However, the setting aside of arable land, as in other parts of the European Union because of agricultural overproduction (Grounds for Choice, 1992), is not yet strongly developed in Saxony or the Leipzig–Halle agglomeration. A considerable part of the potentially fallow land is used for growing of renewable resources, especially oil seeds for production of biodiesel fuel, and this trend is continuing. The loss of agricultural land, formerly associated with coal mining, is now associated with the processes of urbanisation and this problem is discussed in more detail later. Over the next 15 to 20 years, the agricultural policy of the State of Saxony involves planned afforestation of an area of 55,000 ha; this is equivalent to 5.7 percent of the currently used agricultural area.

Livestock numbers decreased significantly between 1989 and 1992/1994 (Table 8.3). The number of cattle fell by about 50 percent and the number of pigs by about two thirds. All this happened within two years. An important additional ecological effect has been the decreased production of liquid manure, as discussed later. Cowsheds and machinery, grain and fertilizer stores are now often used for keeping furniture, cars and other goods. Many of the older cowsheds and pigsties are no longer used.

Table 8.2 Utilisation of arable land in Saxony (%)

	1989	*1992*	*1994*
Cereals and oil seeds	53	63	66
Root crops	13	6	4
Forage plants	28	23	14
Vegetables	2	1	0.4
Horticultural and fruit land	4	2	0.9
Fallow land	0	5	12

What are the reasons, or 'driving forces', for these changes in land use and livestock? They appear to include a combination of the following:

(1) Rapid integration of east German agriculture into the West European market with its strong competition;
(2) Liquidation of agricultural cooperatives by law;
(3) Closing of local marketing organisations;
(4) Closing of slaughterhouses, dairies and canning factories;
(5) Building of large supermarkets having delivery contracts with west Germany and Western Europe;
(6) Regulations of the European Union (EU) (e.g. fallow regulations).

The farm structure has been totally changed during recent years (Table 8.4). The total number of recorded farms (enterprises) has increased, especially the number of part-time farms which now account for about half of all farms. Two-thirds of the agricultural land is occupied by corporate bodies (Table 8.5), as compared with individual farmers. On average, corporate bodies occupied 1079 ha of agricultural land in 1994. The average size of a full-time enterprise for an individual farmer was 99 ha, with 44 ha for a part-time enterprise (in 1992). In the future it seems likely that the larger enterprises will predominate and produce food with greater efficiency.

Table 8.3 Decrease of livestock between 1989 and 1992/1994 in Saxony (in 1000)

	Cattle	*Cows*	*Pigs*	*Sows*	*Sheep*
1989	1261	465	1978	174	468
1992	665	259	789	55	148
1994	647	249	611	51	123

Table 8.4 Number of enterprises (farms) in Saxony

	1989	*1992*	*1994*
Total enterprises	955	5246	7697
Full-time enterprises	955	2668	3538
Part-time enterprises	0	2578	4159

Table 8.5 Portion of land according to land tenure in Saxony (%)

	1989	*1992*	*1994*
Corporate bodies			
Agricultural cooperatives	99.7	0.0	0.0
Others	0.0	72.7	66.3
Individuals	0.3	27.3	33.7

There is a marked tendency towards greater specialisation within agricultural enterprises (Table 8.6). Large enterprises have specialised in commercial farming (mostly crop production), while the number of specialized livestock farms is increasing but remains small. Individually-occupied farms are often 'mixed farms' (crop production plus livestock), although their numbers are falling. The number of market gardening and fruit farming enterprises in 1994 was smaller than in 1989.

For a variety of reasons peasant farms, which were customary in Saxony 40 years ago, are now few in number. The reasons for this trend include:

(1) Children of former peasant farmers have chosen professions other than agriculture;
(2) Former peasant farms were too small to support a family;
(3) Farm buildings were out of date, if they existed at all.

The new full-time enterprises of individual farmers are often tenanted farms run with a part of their own land; some of the occupiers of these tenanted farms live in western Germany and have become so called 'sidewalk farmers'.

These fundamental changes in production and farm structure have resulted in an unprecedented loss of jobs and redundancy of farm labour. In 1989, 180,000 people worked in agriculture and horticulture in Saxony; now only 44,000 are left. The number of workers on cooperative farms per 100 ha was 15.7 in 1989 but only 4 in 1992. But we must also bear in mind that in 1989 craftsmen and services were integrated within agricultural cooperatives; most of the these jobs no longer exist. The number of workers in agriculture in Saxony will continue to decrease, possibly to only 30,000 or 3.3 per 100 ha.

ECOLOGICAL CONSEQUENCES OF AGRICULTURAL RESTRUCTURING WITHIN RURAL AREAS OF THE LEIPZIG–HALLE–BITTERFELD AGGLOMERATION

Agricultural and forestry landscapes function as large accumulators, filters and transformers (regulators) of materials and energy vital to human society. In the recent past, production has been the main function of the rural landscape, including the maximization of yields, and regulatory functions which

Table 8.6 Type of agricultural enterprise farms in Saxony (1992/1994)

	1992		1994	
	Number	*%*	*Number*	*%*
Total	5794	100.0	7697	100.0
Commercial farming (grain, sugarbeet)	535	9.2	2164	28.1
Livestock farming	80	1.4	154	2.0
Mixed farming	4027	69.5	3877	50.4
Market gardening, Fruit farming	1152	19.9	1502	19.5

145

are vital in the long term have often been severely damaged. Therefore, the top priority of ecological activity today is to maintain and restore the regulatory functions of landscapes.

The protection of groundwater from any kind of contamination, even outside those areas designated for gathering drinking water, is the principal concern. Because of over-fertilization with nitrogen, the nitrogen balance has been more than +110 kg N per hectare per year in Saxony and Saxony–Anhalt (Table 8.7). Consequently, groundwater is endangered by NO_3 contamination over large areas; in particular, districts containing large cowsheds and pigsties have had high positive N balances (Figure 8.1, Nolte *et al.*, 1991).

The mean NO_3 content over 12 years in the percolated water collected in lysimeters in Brandis, east of Leipzig, which has a common regional practice of fertilization, is twice or three times higher than the EU norm of 50 mg/l NO_3 from sandy loess soils, depending on the density of the B-horizons and local geology. In percolated water from a loess soil, the NO_3 content was 50 mg/l NO_3. Because there are many smaller waterworks in the region, more attention is needed for the protection of ground water resources (Knappe *et al.*, 1993). Thus, after changing the rotation of crops on the lysimeters (to green fallow land without fertilization), and without organic fertilization of the loess 'sol lessivé' (Lößparabraunerde), the NO_3 content was only 10 mg/l in the percolated water in spring 1994 (Knappe and Keese, 1995).

Since 1990 the input of nitrogen has decreased as a result of reducing the number of cattle by one half and that of pigs by two thirds. Presuming equal chemical fertilization and an equal output of yield as in 1989, the amounts of the nitrogen balance have decreased through reduced production of liquid manure (Figure 8.1). The danger of groundwater contamination over large areas is now less than in the past. But we must bear in mind that over-fertilization in the past can cause a further increase in the NO_3 content in groundwater. Locally, high pollution by liquid manure may also be expected in the future. During recent years mineral nitrogen fertilization has also declined. As a result of the decreased nitrogen input, the NO_3-N content at

Table 8.7 Surplus of nitrogen for Saxony–Anhalt and Saxony (kg N per ha of agricultural land – 1989). Data of the Statistisches Jahrbuch der Land-, Forst- u. Nahrungs-güterwirtschaft d. DDR, 1990

	Supply in kg N per ha				Production of biomass (corn crop units/ha)	Consumption of N by vegetation (kg N/ha)	Balance (kg N/ha)
	Artificial fertilizer	Organic fertilizer	Precipitation	Total			
Saxony Anhalt	128.7	57.6	20.0	206.3	45.5	91.0	+115.3
Saxony	130.9	72.1	20.0	223.0	56.4	112.8	+110.2

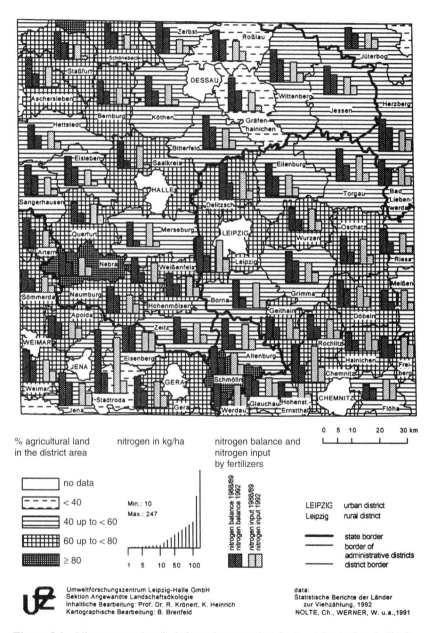

Figure 8.1 Nitrogen surplus (in kg) per hectare of agricultural land of each district 1988/89 and 1992

0–60 cm depth has decreased in Saxony from 120 kg per ha in autumn 1990 to 76 kg in autumn 1993 (Sächsischer Agrarbericht, 1994). The reduced danger of nitrogen contamination of soil, groundwater and surface water is also a result of promoting and supporting 'integrated cultivation', which means that only the quantity of manure consumed by plants is added to the soil. Farms that participate in integrated cultivation receive grants from the EU and the State of Saxony.

Only modest success has been achieved in control of soil erosion. Fields tend to be too large: 50–100 ha fields, with a length of several hundred metres, are common. In the region of Leipzig–Halle the hilly loess districts are especially endangered by water erosion. Because of the lack of protective meadows in valleys, eroded soil is often transported into rivers and lakes. Water erosion contributes to contamination of running water by phosphates and so consequently contributes to eutrophication of the water. On the flat plains in the north and west of the region there is a danger of wind erosion; here there are few hedges to act as wind breaks.

The loess agricultural landscape is, in wide areas, nearly free of forests and trees – except in river valleys and some remaining forests in the south and east of the Leipzig–Halle region – and is of low landscape diversity. The proportion of woodland in the district of Halle is around 13 percent, with 15 percent in the district of Leipzig, thus offering few habitats for wildlife, plants and animals (Table 8.8) (Reichhoff, 1988).

Establishing a higher landscape diversity is of great importance, especially for protection and renaturalisation of natural biotopes typical for the landscape, and for improved aesthetic value in outdoor recreation. For this reason, biotopes need to be created, as a basic framework, in river valleys and the remaining forests so that the degree of naturalness in agricultural landscapes can be increased. However, restoration of the biotic regulatory function cannot be restricted to the natural parts of the landscape but needs to be seen in relation to the total region. Created biotopes should be established in such a way that they can simultaneously function in erosion control.

Table 8.8 Percentages of habitats for plants and animals to be preserved within the loess region

Area (km²)	0–5%	5–10%	10–20%	>20%
to be preserved for plants	6015	2625	371	790
%	61	27	4	8
to be preserved for animals	2625	4105	2390	790
%	26	42	24	8

(according to Reichhoff, 1988, in Stern, 1990)

CHANGES TO LAND USE BY MINING OF BROWN COAL

Mining of brown coal is a problem specific for the Leipzig–Halle agglomeration. About 471 km² have been claimed for such mining (Table 8.9) but only 47 percent of the mined area is given back to agriculture, forestry, water bodies and other uses after reclamation and amelioration. After reclamation the agriculturally-used area covers a distinctly smaller proportion of the land, while the forest area rises (Table 8.10). Table 8.10 shows only an intermediate stage (1991) in the development of part of the Leipzig agglomeration. After completion of the reclamation work, it is planned that 40 percent of the entire

Table 8.9 Brown coal mining in the Leipzig–Halle agglomeration

Region	Mining area (km²)	Reclaimed area (%)	Output of coal 1989 (mill t)	Output of coal 1993 (mill t)	Output of coal total (mill t)	Settlement demolition (number)	Settlement demolition (inhabitants)
Bitterfeld	119.5	45.9	23.6	1.8	970	16	5858
Borna	180.2	47.3	52.3	16.5	3200	67	22492
Zeitz-Weißenfels	54.9	48.5	15.5	7.6	520	12	4718
Geiseltal	52.0	43.5	5.9	1.3	1430	19	11587
Halle	24.7	80.6	5.4	0.0	200	0	0
Nachterstedt	19.4	32.4	0.7	0.0	253	5	3157
Röblingen	14.6	41.6	2.0	0.8	70	0	0
Harbke	5.6	29.9	0.3	0.0	70	0	0
Total	470.9	47.4	105.7	28.0	6713	119	47812

(from Berkner, 1995)

Table 8.10 Cumulative land balance of brown coal mining within the Leipzig region (in ha and %) (from Sächsischer Agrarbericht, 1992)

	Leipzig region	
Land-use category	ha	%
Areas used by mining	25,471	100.0
Agricultural land	21,364	83.9
Forest	2286	9.0
Water	58	0.2
Others	1763	6.9
Reclaimed areas	9368	100.0
Agricultural land	5245	56.0
Forest	2992	31.9
Water	176	1.9
Others	955	10.2

mining area will be used as water bodies, 35 percent for forest, 20 percent for agriculture, and 5 percent will be left for miscellaneous use.

The land reclaimed for agriculture is suitable mainly for growing grain and forage crops. But the tendency of the soil to compress and saturate produces greater variation in yield from year to year compared with undisturbed soil. For this reason the land is particularly subject to rotational fallow, while the intention to use it for additional afforestation has been pointed out already.

Brown coal mining has decreased sharply in the last few years. In the Leipzig–Halle agglomeration, 105.6 million tons of coal were mined in 1989. By 1990 extraction was only 80 million tons; this decreased to 36.3 million tons in 1992 (MIBRAG, 1993; übersichtskarte, 1993) and to 17.4 million tons in 1994 (press report). In future coal will be used chiefly by two new power stations; the old power stations have already been closed or will be closed in the near future.

ECOLOGICAL CONSEQUENCES OF MINING BROWN COAL

We must distinguish between the indirect and direct ecological consequences of mining brown coal. For example, indirect consequences include pollution as a result of burning brown coal in the region, while one of the direct consequences is the change in water balance of the region. On pollution, within the agglomeration of Leipzig–Halle, enormous amounts of calciferous brown coal ash has been deposited on agricultural land in the last 100 years. In the last few years the emission of dust has declined, partly because power stations and chemical plants have been closed down and partly because more natural gas as a source of energy has been imported. According to TÜV Rheinland (1991), the emission of dust has changed as follows:

1988	340,000 t/a
1990	200,000 t/a
1991	157,000 t/a

The Centre for Environmental Research has investigated the question of contamination by pollutants exceeding acceptable thresholds in the soil and the vegetation. Schädlich and Schüürmann (1993) investigated the Ap horizons at 65 points spread over 7000 km². The calcium content and the pH index were found to rise significantly in the lee of sources of pollution, but the absolute amount depends on soil type. The investigation of forage plants by Peklo and Niehus (1993) has shown that the heavy metal content is also lower than the reference data for forage plants. The heavy metal contamination of arable land as a result of air pollution and the emission of dust (ash) is obviously of smaller significance. Greater heavy metal contamination is found locally in city soils and in the soils of flood plains. Heavy metal contamination of the floodplains is not so high as to necessitate withdrawal of land from agricultural production. Below the chemical plants of Bitterfeld/Wolfen, waste water from the production process transports noxious organic substances onto the river flood plain. These pollutants reach a concentration which makes it necessary to exclude some areas from agricultural exploitation (Villwock and Lauer, 1994).

Looking now at the water balance of the region, according to Berkner (1995) the groundwater level in the Leipzig–Halle agglomeration has been lowered after brown coal mining over an area of 1,100 km², in parts by up to 100 m. About 10,000 million m³ of water will be needed to refill the subsidence craters and fill up (flood) the worked-out opencast mines. On the other hand, because of the increasing number of open bodies of water and woodland, the evaporation rate is higher than it used to be on former farmland. The estimated runoff, before mining activities, reached approximately 130 mm m⁻². On the landscapes created by open-cast mining, the runoff will only be 10–30 mm m⁻² on average and this will only happen towards the middle of the next century. Nevertheless, immense ecological problems are already resulting as the rising groundwater level increasingly reaches old waste dumps: the danger of entry of pollutants into the groundwater is rising. Because of depleted groundwater layers, the groundwater locally has risen to a higher level than before mining activities; one result is water entering cellars of many houses.

The reshaping of the landscape following opencast mining has positive effects for recreation and nature protection. Several lakes resulting from the flooding of old workings now function as recreational areas, and cycle and footpaths have been built. They are widely accepted by the population. More lakes will follow. The two biggest lakes will have surface areas of 26 km² and 18 km². In several places natural secondary biotopes containing rare plants and animals have emerged. At the largest of the existing opencast lakes (6.3 km²), in the east of Bitterfeld, some 10,000 wild geese have been counted on their passage to the south in late autumn. Instead of intrusive mining activities on the very edge of the cities of Leipzig and Halle, with heavy blowing of dust from uncovered soils, a varied landscape is in the process of creation, considerably improving the attractiveness of the Leipzig–Halle agglomeration.

LAND-USE CHANGES UNDER SUBURBANISATION

In the old West German federal states, settlements cover 12 percent of the total area. In the new East German states, the proportion is only 7 percent but this area is rapidly expanding through the construction of new settlements. The map of the *approved new settlement area* by law (Figure 8.2, including commercial, industrial and dwelling areas, as well as areas for mixed and special use) shows considerable spatial differentiation. Within the agglomeration of Leipzig–Halle, the city region of Leipzig is the dominant focus of development. By the end of 1992 land-use change for 49 km² had been approved by the regional planning administration for the city region of

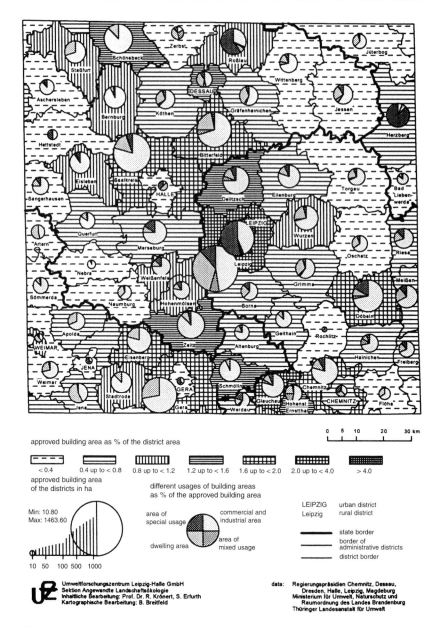

Figure 8.2   Approved new settlement areas for the district (June 1994)

Leipzig (the approved area by law shown in Figure 8.2 is smaller by 50 percent), against the 30.3 km² for the total city area. This comparison of areas clearly shows the beginnings of an expected process of suburbanisation.

The process of suburbanisation is a particular feature of the city region of Leipzig (Regionalplan, 1995). The construction of *commercial facilities* has been carried out most quickly, especially for supermarkets (Table 8.11). The attractive power of motorways on retailers and wholesalers shows at the intersection of the A14 and new B2 highways. Here, on the site of the old Leipzig–Mockau airport, the buildings of the new Leipzig Fair are being erected at a hectic speed so as to out-compete other cities (e.g. Munich). The mailorder firm Quelle, one of the biggest German mailorder firms, has begun to deliver goods to customers from the site. The Fair company, as well as Quelle, intend to open up new markets in Eastern Europe. Nearby a big wholesale centre for fruit and vegetables has opened.

A revival of *housing construction*, especially construction of one-family detached and terraced houses, can be clearly seen. Nevertheless, the number of newly constructed flats is still far below the figures of 1989, when whole quarters were built on the outskirts of Leipzig with prefabricated concrete components. Leipzig is very restricted in its development because of a lack of space and so housing construction takes place outside the city's administrative area. The newly developed housing has individual plots of land covering an area of approximately 350 to 500 m². Ground prices are between 150 and 200 DM/m² if development and sale are the responsibility of the communities. Higher-value building land exists in the city's outskirts; here the land was divided into small plots before the Second World War but has only partially been built on, being used mostly as gardens and week-end properties with only summer-houses or bungalows. Prices of between 250 and 350 DM/m² are paid for plots of approximately 1000 m², and far more in preferred locations. Suburbanisation also shows in migration figures. In 1994 19,102 inhabitants emigrated, whereas only 13,097 migrated into Leipzig. Forty percent of the emigrants moved to the suburban areas of Leipzig.

Leipzig is increasingly developing as a centre of *banking and services*. The centre of the city is the preferred location for banks and offices, as well as

Table 8.11 Newly opened supermarkets in the city region of Leipzig

Supermarkets	Place in relation to Leipzig	Opened	Sales area (m²)
Saale Park	Günthersdorf, west	1991	88,400
Sachsen Park	Seehausen, north	1992	62,500
Globus EKZ	Wachau, south	1993	17,300
Löwen Center	Bienitz, west	1993	29,400
Pösna Park	Großpösna, southeast	1993	32,500
Paunsdorf Center	Leipzig, east	1994	70,000

From: Bundesforschungsanstalt für Landeskunde und Raumordnung, Arbeitspapiere, 7/1995

hotels. A secondary location has been developed in Schkeuditz, near the airport, and the Schkeuditz motorway interchange.

There is a wide range of plots of land suitable for *industrial and commercial* development in the city region and district (Bezirk) of Leipzig, covering 118 industrial and commercial areas. These areas concentrate along a number of main traffic arteries and some district towns (Kreisstädte) (Baulandkatalog, 1994). Only time will tell whether the expectations of a flourishing business life in the new industrial and commercial centres will be fulfilled.

Currently, extension and improvement of the *transport* system has claimed only a small amount of arable land. To begin with, existing traffic routes were repaired and reconstructed, but now there are new development projects, such as the 6-lane extension of the A9 Nuremberg–Berlin motorway, the A14 between the Schkeuditz motorway interchange and the motorway exit at Engelsdorf in the north of Leipzig, as well as construction of the A140 which functions as the outer ringroad around Leipzig. In addition, there are plans for construction of a motorway between Leipzig and Chemnitz (A83) and construction of the A14 motorway between Halle and Magdeburg has begun. There are other plans for a South Harz motorway (Goettingen–Halle) and a North Harz motorway (Goslar–Bernburg–Dessau), which are to directly connect the Leipzig–Halle agglomeration with western Germany. An Intercity railway line from Erfurt–Halle/Leipzig to Berlin is planned and a city train line within Leipzig–Halle will be added. The airport for Leipzig–Halle will get another runway to the north of the A14. Nearby a large freight centre is under construction.

Because of suburbanisation, fragmentation of the landscape will increase together with an increase in environmental pollution caused by traffic. It is remarkable that the pressure of urbanisation finds no resistance from landowners or agricultural administrators. Many landowners have willingly sold land given to them after the break-up of the agricultural cooperatives. The agricultural administrators expect that decrease in arable land will help to reduce the area used for growing cereals and thus meet the limits placed by the EU on growing cereals and oilseeds. The negative ecological consequences of urbanisation are not taken into consideration and the important regulatory functions of the open landscape are lost. The following section, for example, discusses the impact of land-use changes for the regional water balance.

LAND-USE CHANGE AND THE REGIONAL WATER BALANCE

The agglomeration of Leipzig–Halle is situated in the lee of the Harz Mountains. Consequently, the western part of the region is very dry and belongs to the Central German Dry Region. Mean annual precipitation rises from 450 mm in the northwest to 700 mm in the southeast. The mean potential evapotranspiration in the region is approximately 550 mm. Under these climatic conditions the field capacity of soils, the land cover and the distance between groundwater and the surface are all important influences on evapotranspiration and thus on percolation and runoff. It is important to calculate evapotranspiration and runoff because numerous geochemical processes, including water erosion, depend not as much on the absolute precipitation figures as on the water which is percolating, interflowing and running off in a landscape.

With the help of a model (RASTER) using values of precipitation, land cover, soil type, groundwater level and potential evapotranspiration, the regional authority for Environmental Protection of the State of Saxony–Anhalt has calculated data for evapotranspiration and runoff for grid cells of 500×500 m. These data have been aggregated by the author for landscape units and the result shown in Figure 8.3. The relationship between evapotranspiration and runoff in the landscape units is found to differ considerably, as explained further in Table 8.12 for two transects running in an west–east direction. Evapotranspiration is reduced in cities with their high proportion of urban land use while in flood plain landscapes it is increased because of high groundwater levels. In landscapes created by open-pit mining, with a high proportion of their area uncovered by vegetation, evapotranspiration is somewhat lower than in adjacent agricultural landscapes. With a rising percentage of woodland within a landscape unit, evapotranspiration increases and runoff decreases. Large-scale land-use changes will alter basic figures for the regional water balance and, as previously mentioned, in the agglomeration of Leipzig–Halle this will be the case in mining areas because of the increasing number of water bodies and the growing proportion of woodland. The increasing area of urbanised land in the cities will be of local importance.

ECOLOGICAL ACTIONS TO ENSURE MULTIPLE USE OF LANDSCAPES IN THE LEIPZIG–HALLE REGION

De Groot (1992) proposed four classes of function for natural landscapes which are also valid from human-influenced landscapes. These are: *regulatory, carrier, production* and *information* functions. Each landscape has to fulfil one or several functions at the same time from each of the four classes. On agricultural areas, for example, we find cultivated plants (carrier function) for food, forage and renewable resources (production function). The organisation and size of fields points to operating structures (information function). At the same time, the type of land cover influences the microclimate, amount of water needed by plants, runoff, percolation, formation of groundwater resources, and soil erosion (regulatory function).

However, determining the functions to be fulfilled by a landscape is a complex process. For instance, in a society oriented towards short-term profits, carrier and production functions of the landscape are of immediate importance; information functions tend to be used intuitively rather than consciously, while regulatory functions are often ignored. Indeed nature's role

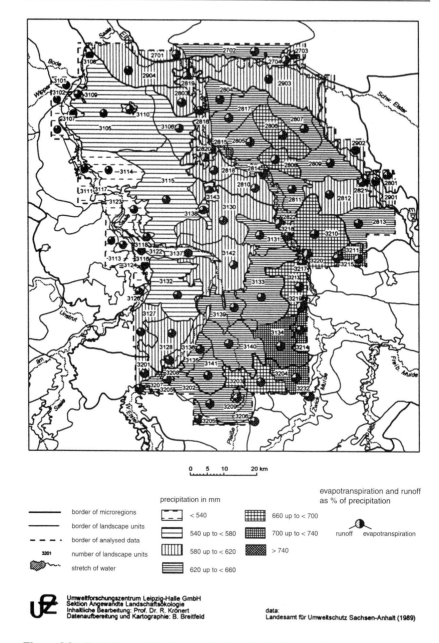

Figure 8.3 Basic figures for the water balance in landscape units

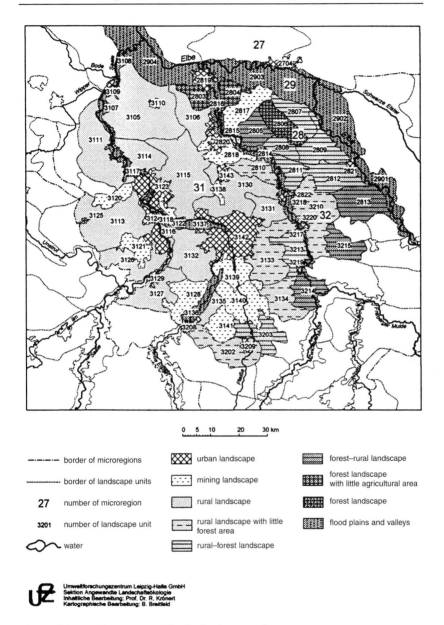

Umweltforschungszentrum Leipzig-Halle GmbH
Sektion Angewandte Landschaftsökologie
Inhaltliche Bearbeitung: Prof. Dr. R. Krönert
Kartographische Bearbeitung: B. Breitfeld

Figure 8.4   Land-cover types for the landscape units

Table 8.12 Basic figures for the water balance in selected landscape units

Area	Number	Precipitation (mm)	Evapotranspiration (mm)	Runoff (mm)	Evapotranspiration (%)	Runoff (%)
Querfurter Platte	3113	524	444	80	85	15
Halle	3123	537	397	140	74	26
Leipzig–Schkeuditzer Elstertal	3137	559	512	47	92	8
Leipzig	3142	598	407	191	68	32
Zwenkau–Espenhainer Bergbaugebiet	3139	621	463	158	75	25
Leipzig–Naunhofer Land	3133	653	493	160	76	24
Grimma–Brandiser Porphyrhügelland	3213	676	508	168	75	25
Köthener Ackerland	3105	540	443	97	82	18
Quellendorf–Thalheimer Ackerland	3106	577	414	163	72	28
Mosigkauer Heide	2803	590	473	117	80	20
Wolfen–Bitterfeld	2820	592	349	243	59	41
Jeßnitz–Dessauer Muldental	2816	610	500	110	82	18
Oranienbaumer Heide	2804	627	513	114	82	18
Gräfenhainichen–Muldensteiner Bergbaugebiet	2817	625	481	144	77	23
Westliche Dübener Heide	2805	641	479	162	75	25
Zentrale Dübener Heide	2806	684	527	157	77	23
Authausener Platte	2808	686	470	216	69	31

in both regeneration and regulation of the landscape are often taken for granted, as shown by the parlous state of the environment. Even today it is often erroneously assumed that agriculture and forestry, when managed according to 'reasonable agricultural usage', will achieve landscape protection on 85 percent of the State's territory; also it is assumed that this can be achieved without cost and that the regulatory functions of the landscape can be maintained without further human intervention.

Neef (1967) argued that natural and constructed landscape functions form a continuum: the landscape consists of urban as well as 'open' landscapes. In the present context, all four landscape classes have to be taken into consideration, covering urban, agricultural and mining. Moreover, there is no value-free judgement as to which functions, in relation to a landscape type, are to be expected and promoted: this depends on the existing social model, guiding principles of development, needs of the people, and their potentialities in transforming guiding ideas into action (Krönert, 1995).

In *urban landscapes*, especially in cities, the most important functions are those directly concerned with life and activities of the people. It is particularly important to offer space for dwellings, work, services and communications, as well as recreation, as basic life functions (i.e. carrier functions for human dwellings, settlement and recreation). Simultaneously, all of the regulatory functions which control the city's climate and keep the chemical composition of the atmosphere within definite limits are of importance. These functions make possible the accumulation and recycling of wastes without contaminating soil and water, and contribute to regulation of runoff and flood protection. Within the natural production function, the production of oxygen is of special significance. If these regulatory functions are to be fulfilled by natural processes, towns and cities must contain as much vegetation and 'open' land as possible. Also, the flood plains of rivers which run through cities must be 'renatured' or maintained in a natural condition. The outcome should be multiple use of the city landscape, including functional overlapping as well as functional mixing (i.e. functional coexistence). Alternative strategies include expensive technical recycling, avoidance of waste and waste products, and using the surroundings of the city to dispose of municipal wastes or where the citizen can escape for recreation. It is not so much a matter of choice between alternatives as forming a balance amongst them. Nevertheless, the restoration of the natural regulatory potential of the urban landscape should be paid more attention.

In *agricultural landscapes*, up to now agricultural production functions have dominated, as have forestry production functions in forest landscapes. Here, almost all regulatory functions have importance, although which functions are locally and regionally emphasised depends on the decisions of the land user or society. In these types of landscape the regulatory function is under considerable pressure, especially under nitrate contamination of groundwater and soil erosion in hill country, with consequences for other functions of the landscape. For instance, groundwater from the agricultural and forest areas is used for urban consumption, so that water protection areas have to be set up. Also, parts of the open landscape have to be placed under the control of nature and landscape protection agencies in order to preserve habitats and landscapes suitable for recreational use. For sustainable functioning of the agricultural landscape, preservation and restoration of the regulatory potential is a main task for comprehensive environmental protection.

In *mining landscapes*, the extraction of raw material (energy resources, building materials, mineral ore) as a production function has priority as long as mining is active. Even so, the regulatory functions of the landscape, largely destroyed by opencast mining of brown coal, must be restored. Here, a combination of production function, recreation function and nature protection function of the landscape is necessary.

CONCLUSION

All regions in eastern Germany experienced the break up of economic structures after reunification in 1990, with consequences for land use and land cover. In rural areas, farm structures have been changed completely under reprivatisation, although large farm units – between 100 and 1000 ha – still predominate. Livestock numbers have decreased dramatically and the danger of contamination of ground water by NO_3 has decreased as a result of reduced output of liquid manure. Cereal growing has increased, whereas growing of fodder crops, potatoes and sugar beet has decreased to a great extent. Due to EU directives, a part of the arable land now lies fallow, although on fertile soils the 'industrial crop' of oilseed rape is permitted on 'set-aside' land to produce biodiesel fuel.

Within urban regions, a rapid process of suburbanization is occurring with irretrievable loss of agricultural land balanced by an increase in trading, business and residential areas. In urban regions car traffic has increased drastically.

Within the agglomeration of Leipzig–Halle, opencast mining has destroyed the rural landscape over wide areas and agriculture has lost many hundreds of square kilometres of land. Villages and settlements housing some ten thousand inhabitants have been destroyed and their lands excavated. At present, mining continues in only three opencast sites to the south of Leipzig. In the reclaimed areas, lakes and forests will dominate in the future: the lakes are already well used for outdoor recreation and their importance will increase after further flooding of mining pits. The landscape that follows on from mining will be useful for protection of biotopes, especially for many threatened species of plants and animals.

Land-use changes have influenced evapotranspiration in the region of Leipzig–Halle to a great extent. With the increase in urban land, formation of groundwater has decreased, while surface runoff has increased. In former mining areas evapotranspiration will grow and runoff will decrease with the formation of large lakes, forests and woodland.

One aim of sustainable landscape development is that each landscape unit should fulfill several functions simultaneously. However, the suitability, capacity and resilience of a landscape is determined by natural–technical conditions and existing landscape processes. Actual demands on the use of a landscape determine which functions and potentials are made use of. Thus the multiple use of a landscape will be characterised as:

(1) A mosaic of the main types of use, with the area of each type fulfilling specific landscape functions and claiming specific potentials.

(2) A combination of main and secondary uses prescribed by law and defined by regulations (e.g. drinking water protection areas, nature and landscape reserves, biosphere reserves, national parks, mining protection areas, areas protected against noise, areas preferably used for agriculture, building land, green belts).

(3) A combination of main, secondary and unmanaged uses for recreation, habitats for wildlife, plants and animals, disposal of substances brought in by air, rainfall runoff, formation of groundwater reserves.

In (1) regional planning, land development plans, and the supervision of construction all guarantee multiple-use of landscapes; here areas will complement each other by functional mixing. In (2) the combination of uses is regulated by law (Ordnungsrecht) and the priorities for landscape functions are established. In this way, direct or indirect landscape planning is carried on. Generally, selected regulatory functions of the landscape are protected. In (3) landscape regulatory functions are developed without management. Here, landscape can be under pressure because of loss of certain regulatory functions – soil erosion, modification of wetlands, over-fertilization, acid rain. Much more attention is needed to protect the regulatory functions of the landscape than at present. It must be shown how land use/land cover can be developed within landscape units to protect and improve regulatory functions, including why the protection of soil structure and quality has at least the same importance as protection of rare biotopes.

REFERENCES

Bach, M. (1987). *Die potentielle Nitratbelastung des Sickerwassers durch die Landwirtschaft der Bundesrepublik Deutschland.* Göttinger Bodenkundliche Berichte, Vol. 93, Göttingen

Bastian, O. and Haase, G. (1992). Zur Kennzeichnung des biotischen Regulationspotentials im Rahmen von Landschaftsdiagnosen. *Zeitschrift für Ökologie und Naturschutz,* 1: 23–34

Baulandkatalog (1994). *Angebotskatalog für Investoren. Gewerbe- und Industriegebiete.* Regierungsbezirk Leipzig im Freistaat Sachsen, Leipzig

Berkner, A. (1995). Der Braunkohlenbergbau in Mitteldeutschland. *Zeitschrift f. d. Erdkundeunterricht*, **47**(4): 151–62

Bundesforschungsanstalt (1995). *Bundesforschungsanstalt für Landeskunde und Raumordnung*. Arbeitspapiere, 7, Bonn

de Groot, R. S. (1992). *Functions of Nature. Evaluation of nature in environmental planning, management and decision making*. Wolters-Noordhoff, Holland

Grounds for Choice (1992). *Four perspectives for rural areas in the European Community*. Netherlands Scientific Council for Government Policy, The Hague

Haase, G., Barsch, H., Hubrisch, H., Mansfield, K. and Schmidt, R. (1991). Naturraumerkundung und Landnutzung. Geochorologische Verfahren zur Analyse, Kartierung und Bewertung von Naturräumen. *Beiträge zur Geographie*, **34**: 1–373

Knappe, S., Moritz, C. and Keese, U. (1993). *Durchsickerungsleistung und N-Austrag unterschiedlicher Böden in Abhängigkeit von der klimatischen Wasserbilanz – Lysimeterstation Brandis – Teilprojekt Forschungsverbund ÖKOR – Erhebung von Basisdaten*. BMFT-Förderkennzeichen 0339419 F/I. UFZ Leipzig–Halle, Leipzig

Knappe, S. and Keese, U. (1995). Lysimeteruntersuchungen – Einfluß geänderter Landnutzung auf Stoffeintrag, transfer und austrag an unterschiedlichen Bodenformen in Lysimetern. In Körschens, M. and Mahn, E.-G. (eds) *Strategien zur Regeneration belasteter Agrarökosysteme des mitteldeutschen Schwarzerdegebietes*. B. G. Teubner, Stuttgart–Leipzig, pp. 499–532

Krönert, R. (1994). Recent changes of land use and their driving factors in the rural area of the Leipzig region (Saxonia) – ecological consequences of the land use changes. In Krönert, R. (ed.) *Analysis of Landscape Dynamics – Driving Factors Related to different Scales*. EUROMAB, Comparisons of Landscape Pattern Dynamics in European Rural Areas. Vol. 3, 1993 Oct. seminar, UFZ Leipzig, Dt. MAB Nationalkomitee, pp. 60–72

Krönert, R. (1995). Ökologischer Handlungsbedarf zur Sicherung der Mehrfachnutzung im Raum Leipzig-Halle. *Zeitschrift für den Erdkundeunterricht*, **47**(4): 163–73

MIBRAG (1993). *Rekultivierung nach dem Bergbau o. J, Bitterfeld*

Neef, E. (1967). *Die theoretischen Grundlagen der Landschaftslehre*. Gotha

Nolte, Ch., Werner, W. *et al.* (1991). *Stickstoff- und Phosphateintrag über diffuse Quellen in Fließgewässer des Elbeeinzugsgebietes im Bereich der ehemaligen DDR*. Umweltforschungsplan BMU, Forsch. vorh. Wasser 10204382 i. A. d. Umweltbundesamtes, Berlin

Peklo, G. and Niehus, B. (1993). *Untersuchungen zur Veränderung luftgebundener Stoffeinträge und ihrer Auswirkungen auf die Vegetation in der Region Leipzig–Halle–Bitterfeld*. Teilprojekt Forschungsverbund ÖKOR – Erhebung von Basisdaten. BMFT-Förderkennzeichen 0339419 F/I. UFZ Leipzig–Halle, Leipzig

Regionalplan Westsachsen (1995). *Entwurf*. Regionaler Planungsverband Westsachsen, Stand Juni 1995, Leipzig

Reichhoff, L. (1988). *Analyse, Diagnose und Prognose der Habitatleistung der Lößagrarlandschaft im Süden der DDR*. Diss. B, Halle

Sächsischer Agrarbericht 1992 (1993). *Freistaat Sachsen*. Staatsministerium f. Landwirtschaft, Ernährung und Forsten, Dresden

Sächsischer Agrarbericht 1994 (1995). *Freistaat Sachsen*. Staatsministerium f. Landwirtschaft, Ernährung und Forsten, Dresden

Schädlich, G. and Schüürmann, G. (1993). *Die Belastung landwirtschaftlich genutzter Böden der Industrieregion Halle–Leipzig*. Teilprojekt ÖKOR – Erhebung von

162

Basisdaten. BMFT-Förderkennzeichen 0339419 F/I. UFZ Leipzig–Halle, Leipzig

Statistische Berichte (1992). *Statistische Berichte der Länder zur Viehzählung.* Berlin

Statistisches Jahrbuch (1990). *Statistisches Jahrbuch der Land-Forst- und Nahrungs-güterwirtschaft der DDR.* Berlin

Stern, K. (1990). *Wirkung der großflächigen Landbewirtschaftung in der DDR auf Flora, Fauna und Boden.* Giessener Abhandl. z. Agrar u. Wirtschaftsforschung d. europäischen Ostens, Duncker & Humboldt, Berlin

TÜV Rheinland, Köln, Inst. f. Umweltschutz u. Energietechnik: Ökologisches Sanierungskonzept Leipzig/Bitterfeld/Halle/Merseburg. Kurzfassung des Abschluß-berichtes, Dez. 1991 i. Auftrage d. Umweltbundesamtes Berlin, Vorhaben-Nr. 10402803

Übersichtskarte Mitteldeutsches Braunkohlenrevier (1993). Stand 1/93, Bitterfeld

Umweltbericht 1991 Freistaat Sachsen (1991). Sächsisches Staatsministerium f. Umwelt und Landesentwicklung, Dresden

Umweltbericht 1994 Freistaat Sachsen (1994). Staatsministerium f. Umwelt u. Landesentwicklung, Dresden

Umweltbericht 1991 des Landes Sachsen-Anhalt (1992). Ministerium für Umwelt und Naturschutz des Landes Sachsen-Anhalt, Magdeburg

Umweltbericht 1993 des Landes Sachsen-Anhalt (1994). Ministerium für Umwelt und Naturschutz des Landes Sachsen-Anhalt, Magdeburg

Villwock, G. and Lauer, M. (1994). Erste Ergebnisse der Untersuchungen zur Schadstoffbelastung der Muldeaue im Kreis Bitterfeld. In Konrad Adenauer Stiftung. *Interne Studien*, **97**: 29–43, Sankt Augustin

CHAPTER 9

LANDSCAPE CHANGES IN ESTONIA – CAUSES, PROCESSES AND CONSEQUENCES

Ü. Mander and H. Palang

INTRODUCTION

The socioeconomic situation in the territory of present-day Estonia has changed at least five times during the past hundred years. The latest change – the reprivatisation of land that began in the early 1990s – is still continuing. Shifting policies and ownership relations have influenced the landscape, changing both land use and landscape diversity. In this chapter, these landscape changes will be analysed.

The analysis will deal with the dynamics of land use and landscape diversity at national and local levels. This approach gives a general overview of the processes behind the landscape changes and enables the reader to comprehend the results of these processes as they are reflected in the landscape. Furthermore, the effect of changes in land use on environmental conditions can be analysed. Finally, the ecological network in Estonia resulting from nature conservation policy and the concept of environmental protection will be described in connection with the idea of a European ecological network – EECONET (see Bischoff and Jongman, 1993; Bennett, 1994).

MATERIAL AND METHODS

Data sources and analysis

The data concerning land use before World War II are derived from official statistics (Cadastre Book of Estonia and Livonia, 1918; Estonian Land Use Cadastre, 1939). Data for the period after World War II are taken from the official Land Cadastre of the Estonian SSR and the Republic of Estonia for different years (Data of the Estonian Land Cadastre, 1990; Land Balance of the Estonian SSR 1996, 1967; Land Cadastre of the Estonian SSR 1945–1985, 1986; Yearbook of the Estonian Land Cadastre 1992, 1993). The assessment of the level of fertilizer use is based on official data from the Ministry of Agriculture.

The impact of land-use changes on the nutrient runoff from catchment areas dominated by agriculture is demonstrated for the Porijõgi River basin (see Mander *et al.*, 1995a). Land-use changes in this area were identified with the help of land-use maps at a scale of 1:10,000 dating from 1987–1990. The present-day land-use pattern has been documented from field work.

The meteorological data (precipitation, air temperature, wind velocity, and humidity measured 6 times a day, as well as daily, monthly and annual mean values) originate from the Tartu-Ülenurme Meteorology Station of the Estonian Meteorology and Hydrology Institute (EMHI). It is located in the northern part of the Porijõgi catchment area. At the Reola hydrological measuring point (EMHI), the daily mean stream discharge (m^3 s^{-1}) for Porijõgi was determined.

The Porijõgi river drainage basin was divided into 8 catchment sub-areas characterized by different land-use structures (Mander *et al.*, 1995a). Since 1987, the water discharge has been measured and water samples have been taken from the closing weirs of each subwatershed once a month. In the South Estonian Laboratory of Environmental Protection (during the period of 1987–1991) and in the laboratory of the Institute of Environmental Protection, Estonian Agricultural University (1991–1995), the BOD_5 value, NH_4-N, NO_2-N, NO_3-N, PO_4-P, total-P, and SO_4 content were analyzed. The total inorganic nitrogen (TIN) was calculated as the sum of NH_4-N, NO_2-N, and NO_3-N. All water analyses were made by applying the COMECON countries' standard methods which are compatible with international methods for the examination of water and waste water quality (APHA, 1981). This paper presents some data on the nutrient dynamics in the whole catchment area of the Porijõgi River.

Map analysis

For the investigation of changes of landscape diversity in Estonia during the past hundred years, 30 test areas (each of 20 km^2) were chosen from the territory of present-day Estonia (Palang, 1994). They are equally distributed in four parts of the country (north, west, central and south Estonia). The availability of maps determined the scale of the research. Landscape changes were traced on four maps dating from different times: (1) a Russian topographic map (1:42,000) that was compiled in 1895–1917, later referred to as dating from 1900; (2) Estonian topographic map sheets (1:50,000) from 1935–39, later referred to as dating from 1935; (3) a land-use map (1:50,000) from 1960; (4) Soviet topographic map sheets (1:50,000) from 1977–90, later referred to as dating from 1989. Land use served as the basis for differentiating landscape units. Linear elements – borders of land-use types, bigger roads, rivers, lakes, streams, main ditches – were considered as borders between landscape contours. The areas and perimeters of these units were measured and, based on these values, several diversity indices were computed – heterogeneity (Shannon–Weaver index H; see Bastian, 1994), evenness (e; after Pielou, 1966), complexity of patches (T), 'curvedness' of patch borders (P) (irregularity, formerly also referred to as 'cavityness'; Mander *et al.*, 1995b); edge index (I) – using the following formulae:

$$H = -\sum \{(n_i/N)*\log(n_i/N)\} \qquad (9.1)$$
$$e = H/\log S \qquad (9.2)$$
$$T = 4\pi A/p^2 \qquad (9.3)$$
$$P = p/\{2(\pi A)^{\frac{1}{2}}\} \qquad (9.4)$$
$$I = \sum (p_i)/A \qquad (9.5)$$

where n_i = number of units of i-th land-use type,
N = total number of land-use units,
S = number of land-use types,
A = area of land-use unit (hectare),
p = perimeter of land-use unit (m).

The indices stand for the different types of landscape diversity. While the patch complexity, as well as the edge index, characterise the topological diversity of the landscape, heterogeneity and evenness describe the distribution of the different types of landscape units. The results of such an investigation could be used for planning ecological networks.

DRIVING FORCES OF LAND-USE CHANGES

Physical–geographical conditions

There are two main boundaries which split Estonian territory into regions with different land-use patterns. First, according to Varep (1964), the upper limit of local glacial lakes divide Estonia into two parts – Lower and Upper Estonia (see Figure 9.1). Lower Estonia, which at one time formed the bottom of the sea or some local glacial lakes, is a low plain with large bogs and forests. Upper Estonia, on the contrary, has never been flooded by local lakes and, therefore, the landscape pattern is much more of a mosaic, with different glacial, glaciofluvial, and glaciolimnic landforms, such as drumlins, eskers and kames, prevailing. Especially in southeastern Estonia, the landscape is very diverse. This division is important with respect to the age of the landscapes. The ice cover left Upper Estonia approximately 12,000 years ago whereas landscape development in Lower Estonia was able to begin only some 7000–9000 years ago.

Second, the border between the Ordovician/Silurian and Devonian bedrock formations influences both soils and vegetation. North of this border, Ordovician and Silurian limestones make the soils more alkaline, whereas in Southern Estonia, where Devonian sandstones occur, acidic soils predominate.

The third border that divides Estonia is a climatic–biogeographical one, first described by Lippmaa (1935). It separates more maritime western Estonia from more continental eastern Estonia. Between these two regions there lies a transition zone in the form of a large belt of forests and bogs stretching from the north to the very southwest (Figure 9.1). Furthermore, the vegetation characteristics (e.g. the dominant species in raised bogs) varies between western and eastern Estonia.

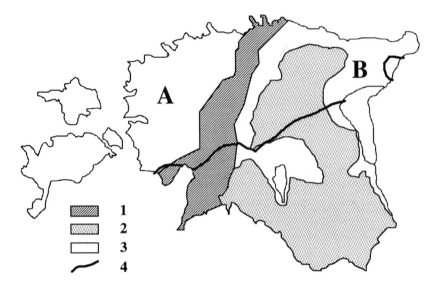

Figure 9.1 Main landscape features of Estonia. 1. climatic–biogegraphical transition zone between more maritime (A) and continental (B) areas (*Estonia intermedia* (Lippmaa, 1935)); 2. Upper Estonia; 3. Lower Estonia (has been flooded by local glacial lakes; Varep, 1964); 4. border between Silurian limestone (northwards) and Devonian sandstone (southwards) formations

Political issues

Land reforms in the first half of the century

The first land reform of the new Republic of Estonia was proclaimed on October 10th 1919 and was implemented in subsequent years. It strongly influenced the structure of land ownership. At the very beginning of the century, most of the land belonged to Baltic–German landlords. According to the reform, land was nationalised and later distributed to Estonian peasants. The former large estates were divided into small land holdings, which resulted in a mosaic landscape pattern as well as bringing into cultivation much formerly unused land.

Following the first Soviet occupation, one of the first actions was to redistribute the land. The Soviet land reform was announced on July 23rd 1940, only a month after the annexation of the territory. According to this declaration, land was again nationalised. In reality this meant that all land exceeding 30 ha per farm was reallocated to the State Land Reserve to support the creation of new farms. Everybody possessing less than 10 ha of land was supposed to get some more so that the average size of farm could grow to at least 12 ha. In fact, an average of only 10.4 ha was achieved (Kasepalu, 1991).

Altogether, some 50,000 new farms were established during the first year (1940–41). During World War II, occupying German forces returned most of the reallocated land to its former owners.

Soviet times: collectivization, deportations, urbanization and the concentration of agricultural production

In the autumn of 1944, Soviet troops again entered Estonia and a new era in land use began. Initially, the land-use situation of 1940–41 was reestablished as completely as possible. Farmers were made to believe that the land they possessed was theirs without restriction but, starting around 1947, farmers were increasingly pressed to unite into collective farms – kolkhozes. The first kolkhozes in Estonia were formed in 1947, but the majority of the farmers only joined kolkhozes after the great deportation of March 23rd 1949. The following summer (1950), the kolkhozes occupied 87 percent of arable land and employed 82 percent of farmers.

Two great deportations occurred at the beginning of the Soviet regime in Estonia. On June 14th 1941 and March 23rd 1949 respectively, 10,157 and 20,702 persons were deported, leaving the countryside without its land owners and businessmen. The latter deportation, together with collectivization, created a kind of chain reaction that ruined rural life for a decade. The most skillful farmers were deported and those who remained were formed into kolkhozes, the work being done by women and unskilled men. This resulted in a worsening situation bordering on famine, with wretched wages being paid in kind.

Turning now to urbanization, this is one of the factors significantly connected with and influencing the land-use pattern. The 1950s were a time of rapid growth of smaller towns and townships (see Figure 9.2); while the bigger towns obtained their new inhabitants through immigration, the smaller towns mainly gathered 'refugees' from the countryside who were escaping the miserable living conditions of the early kolkhozes. Starting from the 1960s, the situation in the countryside improved, and the concentration of the rural population began. This resulted in central settlements (some for every kolkhoz), which attracted mostly younger people and specialists. Also, new plant and machinery was brought to the countryside which eased farm labour.

The increased concentration of land and agricultural production in Estonia is mirrored by the distribution of animal farms (Figure 9.3). In 1939, the number of animals in farms was equally distributed over the territory, varying from 3 to 9 animals per farm. In 1990, just before the collapse of collectivized agriculture, there were significantly fewer farms and a much higher concentration of animals. On the largest cattle and pig farms, for instance, the number of animals was as high as 2500 and 80,000, respectively. Large farm complexes were concentrated in the eastern part of Estonia as a result of the fertile soils of Upper Estonia (see Figure 9.1) and for political reasons.

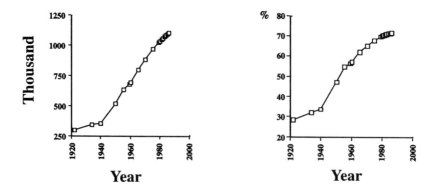

Figure 9.2   Growth of the urban population in Estonia

Current regional policy and re-privatization

In the late 1980s, during the collapse of the Soviet regime, the reprivatisation of land began and, to regulate this process, several legislative procedures were adopted. For instance, the Land Reform Act now controls the return and replacement of land as well as the terms of any financial compensation. The Act applies to individuals whose land was nationalised on June 16th 1940 and, to a limited extent, to legal bodies (including non-profit organizations and the church). Former owners or their descendants are given first priority concerning land claims, except if the land has already been allocated under the 1989 Act on Private Farming. In some specific cases, part of the land will remain in state ownership. Land lots, which are not subject to return or substitution and which are not to remain in state or municipal ownership, may be privatized.

Problems in the privatization process stem from the too generous definition of what constitutes the circle of relatives of former land owners. Also, privatization has resulted in a reduction of agricultural production by more than 30 percent compared with the level of 1991 (see Mander and Palang, 1994). Agricultural production has also been reduced as a result of low prices for farm products as a result of imports from Scandinavian countries, Germany and The Netherlands; a lack of knowledge and experience in marketing has also been to blame.

According to the Yearly Cadastre, the area of agricultural land in Estonia was 1,499,555 ha (32 percent of total area) and the area of arable land was 1,127,824 ha (78 percent of agricultural land) on January 1st, 1995 (Agriculture, 1995). At the same time, there were 983 active agricultural enterprises (Estonian Enterprises Register) and 13,513 private farms (National Land Board). The average size of a private farm was 23.1 ha (Agriculture, 1995). Agricultural enterprises (cooperatives, state farms, and agricultural auxiliary enterprises) have been reorganized from former kolkhozes and

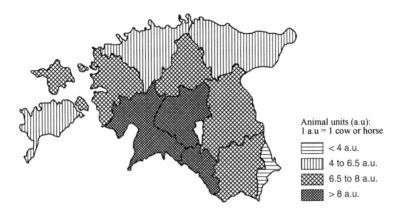

Figure 9.3A   Concentration of agricultural production in Estonia. Average number of animals on farms in 1939

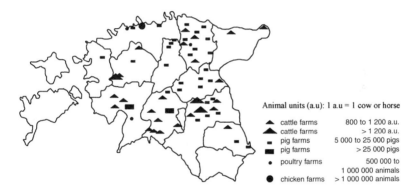

Figure 9.3B   Concentration of agricultural production in Estonia. Location of large animal farms in 1990

sovkhozes. Although the cadastral data show a relatively low decrease in agricultural land, a significant part of formerly used arable land and cultivated grassland is set aside. Likewise, the number of livestock was significantly reduced. For instance, the number of cattle and pigs dropped from 770,000 and 960,000 in 1991 to 420,000 and 470,000 in 1994, respectively (Agriculture, 1995).

Land reclamation

Land reclamation has been one of the most important driving factors in land-use pattern changes. Before the Soviet period, in 1940, the total area of the reclaimed land was 350,074 ha, 44,488 ha of which was drained (Juske *et al.*, 1991).

Under the Soviets this figure nearly doubled, amounting to 731,000 ha in 1991 (Figure 9.4). The increase was primarily achieved by two campaigns: the first took place in the mid-1950s and was brought about mainly by open-ditching; the second began in the mid-1960s with extensive draining activities. Both activities reduced the area of natural grassland. These campaigns culminated in the 1970s with intensive land reclamation and the re-allocation of the resulting land area (in German: *Flurmelioration, Flurbereinigung*). Many ecological elements were eliminated from rural landscapes. However, the share of agricultural land diminished simultaneously: the greater the effort put into gaining new agricultural land, the lower the effort directed at main-taining older land, resulting in an increase in the forest area. This, in turn, led to great disturbances in the nutrient fluxes through rural landscapes.

THE MAIN TRENDS

Changes in the land-use pattern

The situation at the beginning of the century

The land-use situation at the beginning of the twentieth century was deter-mined by the abolition of serfdom in Russia in 1861. After this, every peasant was allowed to buy land and a massive renting and buying of land for perpetuity began as landlords recognized the possibility of financial profit. Mainly marginal areas were redistributed. As the established system of ancient fields remained unchanged, the new farms consisted of either fragmented pieces of land separated by other farms, or narrow long strips. The marginal situation of the new farms also supported the wide distribution of pastures or meadows.

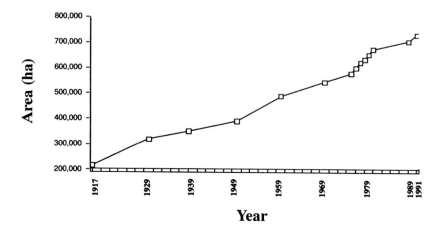

Figure 9.4   Land reclamation in Estonia

172

During the first decades of the twentieth century, land reclamation activities intensified. While in the 1850s open ditching was common in the estates, by the beginning of the new century drainage was used on about 70 percent of the reclaimed land. Between 1897–1918, the Estonian and Livonian Bureau for Land Culture supervised the draining of 366,000 ha, and by 1917 108,000 ha (20,000 ha of meadows, 15,000 ha of drained arable land, the rest in forests) had been treated (Karma, 1959). In 1918, the highest percentage of agricultural land was registered. Even in Virumaa, the county in Estonia where agriculture is least important, the share of agricultural land exceeded 50 percent, while in Saaremaa, the figure was as high as 88.5 percent. These changes were caused by an increase in livestock grazing that, step-by-step, became the most important branch of agriculture, replacing crop growing.

The first period of independence 1919–1940

On October 10th 1919, the Parliament of Estonia nationalised the land formerly belonging to Baltic–German landlords. Immediately, the distribution of nationalised land to peasants began. The 23,023 tenants owning land before the reform maintained their possessions (Kasepalu, 1991). However, the number of persons applying for land increased constantly and by 1924 all nationalised land had been distributed.

A special Homestead Board was created in 1929. Its main task was to occupy unused mineral soil areas as well as wetlands. Due to the market situation, the Government of Estonia supported grain growing in the 1920s, resulting in increased land reclamation activities. Between 1921–1929, 16.7 percent of the arable land and 7.3 percent of the grassland were ditched, while, paradoxically, during the economic crisis of 1929–1933, more than 400 new farms were established. A further 2000 farms were created by 1939, which altogether resulted in 35,000 ha of less-fertile mineral land becoming occupied. Thus during the 20 years of independence, a total of about 350,000 ha were reclaimed (Juske *et al.*, 1991).

Summing up the trends in the land-use pattern, altogether 33,180 new homesteads were established during the first independence period, the total number of farm holdings by 1939 being 139,984 (Kasepalu, 1991). During this period, the area of pasture increased by a factor of 1.36 and that of arable land by 1.44, reaching 1,072,700 and 1,141,800 ha respectively (Eesti, 1993). At the same time, the share of agricultural land diminished slightly, being higher in the western and lower in the eastern counties.

Concentration during Soviet times

In 1942, after the first Soviet year, and the half a year of German occupation, the share of agricultural land had generally diminished, especially in Pärnumaa and Võrumaa. This was due to the consequences of the first deportation and the war. By 1945, after three years of German occupation, the land-use pattern was similar to that of 1939. The share of agricultural land had increased in western counties and slightly decreased in eastern counties.

The most dramatic changes in land use took place in the period between 1945 and 1955: a time of significant change in the social and economic life of Estonia. Most of the agricultural area along the border with the USSR was classified as 'state reserve', which implied the end to all agricultural activities. The entire coastal zone of Estonia (about 3700 km) and many islands lost their agricultural importance (Figure 9.5; Mander and Palang, 1994).

In the concentration of Estonian agriculture during the Soviet period, several stages can be distinguished: first, uniting private farms into kolkhozes between 1947–1951; second, uniting small kolkhozes into larger ones between 1950–51; third, connecting weaker kolkhozes with state farms (sovkhozes) in the late 1950s; fourth, unifying neighbouring kolkhozes in the early 1970s (up to 1976); lastly, the privatization process (i.e. accelerated deconcentration) that began in the late 1980s and has now been stabilized. All of these stages were connected with changes in the land-use pattern.

Summarizing, a wave-like (pendulum-like) shift in the dynamics of the land-use pattern can be observed: after political collapses and during the Soviet period (1944–1991) the share of agricultural land decreased in the western part of Estonia and increased in the eastern part. Opposed to this, during the periods of independence (1919–1940 and 1991–1994), the main increases in agricultural land took place in the western part (Mander *et al.,* 1994b). Overall the land-use changes look like the consequence of a huge 'gravitational' influence from the East.

Dynamics of landscape diversity

The following results must be viewed as preliminary in character because the total area of the test sites represents only 1.5 percent of Estonian territory. On the other hand, the 30 test sites have been chosen from various landscape types and, according to Varep (1964), about 70 percent of landscape types are represented. In addition, the use of maps from different times causes problems when comparing the data; it is possible that measured changes in the landscape pattern reflect only differences in working methods and the interpretations of the map compilers. Nevertheless, maps are often the only source that enables us to trace historical landscapes.

Significant changes have occurred in some test areas, but landscape diversity in Estonia in general has remained stable for the time period under investigation. Within this stability, the following trends of change can be described. As previously argued, this century can be divided into three broad periods of landscape change. The first period was characterized by the process of unification and the growth of evenness in landscape character. During the next period the landscape was divided into smaller units and it became more

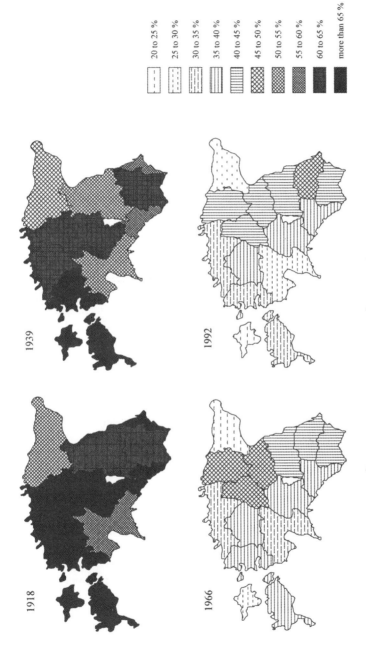

20 to 25 %
25 to 30 %
30 to 35 %
35 to 40 %
40 to 45 %
45 to 50 %
50 to 55 %
55 to 60 %
60 to 65 %
more than 65 %

1939

1992

1918

1966

Figure 9.5 Dynamics of the share of agricultural land in the counties of Estonia 1918–1992

rounded in shape. The last thirty years have brought about the process of polarization and the division of the land into two major groups (fields and forests), while the share of other land-use types has diminished (Rodoman, 1974).

Regional differences can be identified in landscape diversity as well as in the dynamics of diversity. North Estonia is rich in forests, moderately heterogeneous, with relatively large and simple contours and a quite simple landscape structure (Figure 9.6). Its character was preserved from the beginning of the century up to the 1960s, the biggest changes having appeared during the last thirty years. South Estonia has a much more complex landscape structure: the contours are small and with long borders, the area as a whole is uneven, heterogeneous and, in general, returning to its former state after some decades of disturbance (Figure 9.6). Central Estonia is the richest in fields; the landscape pattern is simple and homogeneous and still quite stable. West Estonia is a stable region, with long borders, complex patterns and moderate heterogeneity.

Comparing the rhythms of change in both the land-use pattern and landscape diversity, some similarities can be found. Although the dependence of landscape diversity on land use is not proven, the main tendencies are almost the same. Changes have occurred within the different regions of Estonia, but the landscape diversity of the whole territory of Estonia has

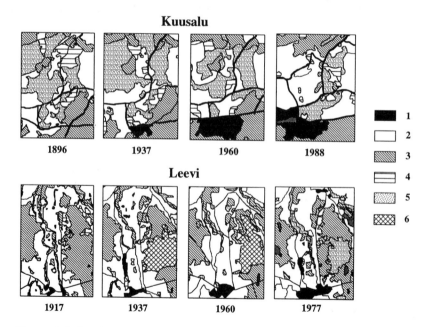

Figure 9.6 Landscape dynamics in Kuusalu, Harfju County, north Estonia in 1896–1988 and in Leevi, Võru County, south Estonia in 1917–1977. 1. settlement, 2. arable land, 3. forest, 4. bush, 5. natural grassland, 6. wetland

176

remained at the same level for the whole time period investigated (see Figure 9.7). It seems as if the inertia of the social system of Estonia, as a rather sparsely populated country, is so powerful that the system as a whole is able to absorb the energy spent by mankind in changing the landscape in its different parts. In other words, land-use marginalization in one region appears to have been compensated by intensification in another.

ENVIRONMENTAL CONSEQUENCES

The main environmental consequences of landscape changes are: (1) the loss of habitats, (2) species extinction and the invasion of new species, (3) the eutrophication of water bodies, (4) groundwater pollution, and (5) an increase in the level of vulnerability of ecosystems.

The *loss of habitats* has been caused mainly by the decrease in natural grassland (Table 9.1) and by land reclamation that has shifted land-use activities from former arable land at relatively high altitudes to marginal areas (former semi-natural grassland, depressions, wooded meadows, riparian and coastal meadows). Meadow communities, often rich in species (690 species, some of them very rare, have been recorded in meadow flora – Külvik 1993) form beautiful patterns in the Estonian countryside, while wooded meadows, with their dense herb layer rich in orchids and single oak, ash and lime trees, are particularly important. Some of the former wooded meadows have been transformed into grassland, others have been afforested, while a considerable number have been overgrown with scrub. This has caused the loss of some species and a number of presently common species will probably become rare of die out in future. Eighty-three plant species are already in danger of extinction (Külvik, 1993). Alvar meadows, the mostly treeless and meadow-like communities on limestone plateaus which are typical ecosystems of Scandinavia, are also threatened. In recent years they have become overgrown with pine and juniper because sheep are no longer grazed in most of these areas. Formerly, sheep effectively controlled the growth of bushes and trees.

In a few decades over 700,000 ha of water-logged meadows, fens and bogs have been drained. As the agricultural use of these areas is not successful, a major part of the land has been covered with young forests and shrubs of low economic value. Fortunately, due to the abundance of all kinds of bogs (about 22 percent of Estonian territory is still covered with bogs – Peterson, 1994), large areas of scientific importance have been maintained.

Loss of species. At present, Estonian flora is made up of approximately 8,600 plant and fungi species, and fauna of approximately 18,500 species (Peterson, 1994). Due to habitat losses, 77 formerly registered plant species (among them, 17 sp. of vascular plants, 21 sp. of bryophytes, 1 sp. of algae, and 38 sp. of macrolichens) have become extinct during the last century (Külvik, 1993). A significant loss (35 percent) of the local nesting ornitho-fauna in a polder territory of South Estonia has been observed because of the

177

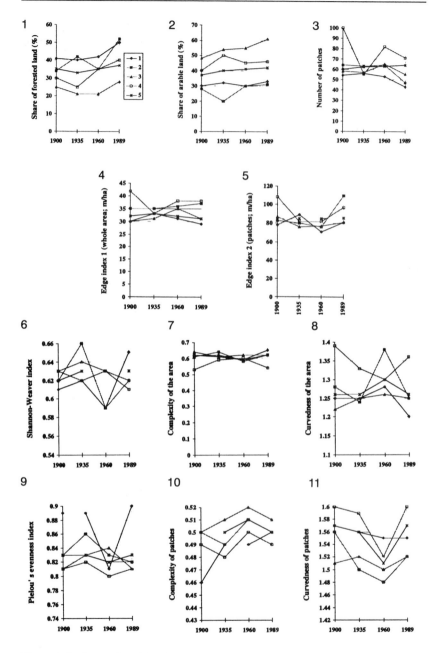

Figure 9.7 Dynamics of landscape diversity parameters in Estonia during the twentieth century, on the base of 30 test areas. 1. north Estonia, 2. west Estonia, 3. central Estonia, 4. south Estonia, 5. whole of Estonia (average values)

Table 9.1 Development of land use in Estonia during the twentieth century (in 1000 ha; after Mander and Palang, 1994 and Agriculture, 1995)

Year	Agricultural land	Arable land	Natural grassland	Forest
1900	2978	1142	1836	796
1918	3092	1135	1792	995
1929	2749	1078	1625	726
1940	2652	1112	1540	873
1945	2455	979	1478	1037
1950	2497	949	1481	1021
1955	2175	968	1385	986
1960	2038	985	994	1462
1065	1851	1030	819	1582
1970	1643	1062	580	1722
1075	1560	1116	443	1793
1980	1500	1134	366	1902
1985	1479	1140	339	1915
1990	1461	1147	311	2012
1995	1450	1144	307	2016

loss of habitat (Mander *et al.*, 1989). In another case study from south Estonian and a Lithuanian moraine landscape, it was discovered that even in isolated woodlots plant species diversity has been influenced by nitrogen from adjacent farms and intensively fertilized fields (Mikk and Mander, 1995).

For many groups of organisms, an invasion of new species and an increase of local populations has been observed. Typical examples are adventitious flora elements and weeds. Another well documented tendency is for an increase in the population size of some mammals (e.g. beaver, wild boar, moose, roe deer, brown bear, wolf, lynx and otter – Küvlik, 1993) and birds (e.g. white stork, great cormorant, mute swan and herring gull). This increase is caused by regional changes in environmental conditions and by landscape changes. For instance, the increase in forested areas and the polarization/ marginalization of the landscape (concentration of agriculture and other economic activities on the one hand; extensification of economic activities in marginal areas on the other hand) have created better living conditions for beaver, brown bear, lynx, otter and wild boar. Also, formerly well regulated hunting played an important role in the increase of roe deer and moose populations. The last three years, however, have brought a significant decrease in the populations of these two species.

Eutrophication of water bodies and groundwater pollution is mainly caused by very intensive use of mineral fertilizers in recent decades, the concentration of agricultural production in big farm complexes, and land reclamation. In central and south Estonia about 60 percent of dug wells and shallow bore wells are contaminated with nitrates (NO_3 concentration > 45 mg/l – Metsur,

1993; Mander *et al.*, 1994a). Also, during and after land reclamation activities, the leaching of nutrients into groundwater and surface water bodies increases significantly.

However, the recent collapse of the collectivized farming system has brought about a significant decrease in the use of mineral fertilizers and land reclamation activities. For example, the mean annual application of mineral and organic fertilizers in Estonia has dropped drastically during the last four to five years: from 0.11, 0.06, 0.93, and 1.03 million tons of N, P, K, and organic fertilizers, respectively, in 1987, to 0.073, 0.047, 0.078, and 0.43 million tons in 1994 (Figure 9.8). As a major consequence of these changes, the nutrient load of rivers, lakes and coastal waters has decreased significantly. Long-term investigations of nutrient cycling carried out in the Porijõgi River

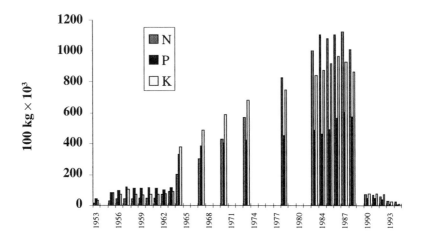

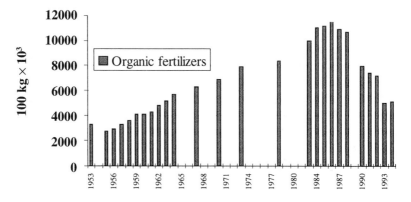

Figure 9.8 Application of fertilizers in Estonia from 1953–1994

catchment, south Estonia, demonstrate a highly significant decrease in both nitrogen and phosphorus losses. The mean annual runoff of TIN, NH_4-N, NO_3-N, and total-P was reduced from 16.0, 1.15, 14.8, and 0.33 kg per ha in 1987 to 2.4, 0.43, 2.0, and 0.22 kg per ha in 1994 (Figure 9.9b). In the same time period, the share of set-aside arable land (fallow), forested areas, semi-natural grassland and wetlands (abandoned drainage areas) increased from 1.7, 40.0, 6.7 and 3.4 percent to 12.5, 44.5, 10.2 and 3.7 percent. Opposed to this trend, the percentage of arable land showed a significant decrease: from 41.8 to 22.5 percent (Figure 9.9a). Besides these significant land-use changes, fertilizer applications in the watershed dropped at the same pace as in all of Estonia.

Increasing rate of vulnerability of ecosystems. This phenomenon is mainly characterized by the appearance of the first signs of damage caused to forests

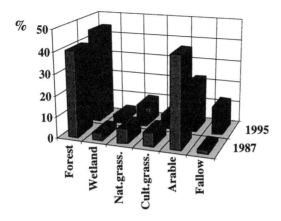

Figure 9.9a   Percentage decrease in arable land

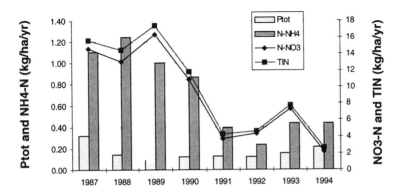

Figure 9.9b   Dynamics in land-use pattern (left) and nutrient runoff (right) in the Porijõgi River catchment, south Estonia. TIN = total inorganic nitrogen

181

and lakes by acid rain. At present, such damage has been detected on about 3200 ha of forest. Similarly, a decrease in both the average age of needles and the radial increment of coniferous trees is probably a result of acidification. Moreover, in the shale-oil industry region of northeast Estonia, alkaline ash deposits from big power plants, cement factories and other industrial sources are disturbing the natural acidic bog ecosystems. One of the consequences of such a disturbance is the significantly lower rate of growth of sphagnum mosses.

ECOLOGICAL NETWORK

A network of ecological compensation areas (an ecological network, see Figure 9.10) can be regarded as a subsystem of the anthropogenic landscape. Such a network counterbalances the impact of anthropogenic infrastructures in the landscape and preserves the main ecological functions of landscapes, namely:

(1) to compensate for and balance all inevitable outputs of human society,
(2) to receive and render harmless all unsuitable wastes from populated areas,
(3) to recycle and regenerate resources,
(4) to provide wildlife refuges and conserve genetic resources,
(5) to serve as a migration tract for biota,

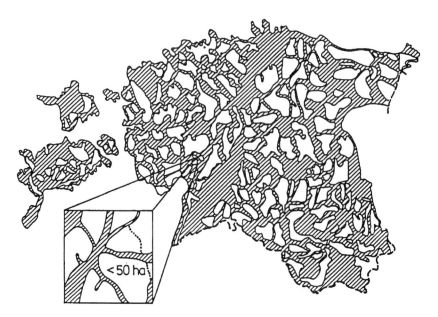

Figure 9.10 Network of ecological compensation areas (ecological network) of Estonia

182

(6) to be a barrier, filter and/or buffer for fluxes of material, energy and organisms in landscapes,

(7) to be a support framework for regional settlements,

(8) to provide recreation areas for people, and

(9) to accumulate material and human-induced energy dispersion (Mander *et al.*, 1995b).

All of these functions are time-dependent and could break down with continuing anthropogenic impacts. Therefore, the ecological network is functional only when combined with additional measures of environmental protection and nature conservation. Ecological compensation areas, combined with areas of intensive human activities, form a strongly polarized (nonbalanced) system that can reduce entropy and increase the self-regulation of a region. However, to level out differences between these two poles (Rodoman, 1974), buffering areas are of great importance.

A network of ecological compensation areas is a hierarchical system with the following levels: (1) core areas, (2) buffer zones of core areas, corridors and stepping stones, and (3) nature development areas that support resources, habitats and species (Baldock *et al.*, 1993). According to the Law on Protected Areas, Species and Natural Monuments, from May 1994 there are five main types of nature protection areas which can be considered as core areas of the ecological network in Estonia (Peterson, 1994; see Table 9.2).

In addition to protected areas, all large forests, wetlands (bogs, swamps, coastal wetlands), lakes, rivers, natural grassland and other large natural communities belong to the ecological network. In some cases ecological farms (31 farms were licensed in 1994) and even former Soviet military bases in forests and bogs can be supporting areas (nature development areas, corridors, stepping stones) for the ecological network.

The size of network components is another feature of the network's hierarchy:

(1) *macro-scale*: large natural core areas (> 1000 sq. km) separated by buffer zones and wide corridors or stepping stone elements (width > 10 km);

(2) *meso-scale*: small core areas (10–1000 sq. km) and connecting corridors between these areas (e.g. natural river valleys, semi-natural recreation areas for local settlements; width 0.1–10 km);

(3) *micro-scale*: small protected habitats, woodlots, wetlands, grassland patches, ponds (< 10 sq. km) and connecting corridors (stream banks, road verges, hedgerows, field verges, ditches; width < 0.1 km).

In Estonia, there is one main 'ecological axis' which represents the macro-scale ecological network: the large forest and bog zone (phytogeographically known as *Estonia intermedia*; see Figure 9.1) extending from Lahemaa National Park in the north to the coast of the Gulf of Riga. Other major axes are not as compact and consist of meso- and micro-scale compensation areas.

Table 9.2 Protected areas in Estonia (after Sepp *et al.*, 1994)

Protected areas	Number	Total area (ha)
National Parks	**4**	**120,940**
State Nature Reserves	**5**	**63,400**
Reserves	**237**	**209,539**
Landscape reserves	77	93,400
Mire reserves	32	96,799
Botanical reserves	50	6800
Botanical–Zoological reserves	16	7100
Ornithological reserves	13	2100
Water reserves	18	2900
Geological reserves	12	440
Nature Park	**1**	**44,200**
Programme Areas	**2**	**1,865,900**
West-Estonian archipelago	1	1,560,000 (land: 404,000 ha)
Biosphere reserve		core zones: 13,600 ha
Pandivere hydrological reserve	1	305,900
Protected Natural Objects	**≈ 3260**	**≈ 8500**
Single elements of landscape (karst areas, waterfalls, outcrops of Paleozoic sediment rock, etc.)	119	600
Erratic boulders and stone fields	264	
Parks	542	6500
Primeval trees and tree groups	537	
Archeological sites	≈ 1800	≈ 500

Ecological compensation areas of all levels comprise about 55 percent of Estonia.

Land-use changes during the twentieth century have significantly influenced the ecological network in Estonia. Except for the core areas, the whole network, especially the semi-natural meadows and forests, serves as a buffering area. For example, during the intensification of agricultural activities the buffering area decreases and it expands when activities are less intensive. Most recent trends allow us to forecast an expansion of the buffer zones of core areas and of nature development areas; this development should support biological and landscape diversity.

CONCLUSIONS

(1) The main trends in Estonian land-use dynamics have been a decrease in agricultural land (from 65 percent in 1918 to 30 percent in 1994) and an increase in forested areas (from 21 to 43 percent).

(2) After recent political changes, the main agricultural activities have been shifted from the western part of Estonia to the eastern part. This pendulum-like movement is caused by the geopolitical location of Estonia.

(3) Land-use changes, the concentration of agricultural production, land amelioration, and the shale-oil based industry in northeastern Estonia, have caused the main ecological disturbances and a great polarization of rural landscapes.

(4) Land reclamation has shifted agricultural activities from the former arable land to marginal areas (former natural grasslands and wetlands); this trend has strongly disturbed the landscapes' nutrient cycles.

(5) The increased use of mineral fertilizer and intensive land reclamation during the last three decades have been essential factors causing eutrophication and groundwater contamination by nitrates. However, the end of the collectivized farming system in recent years has led to a significant reduction of the use of mineral fertilizer and land reclamation.

(6) The concentration of agricultural activities has been a main reason for the eutrophication of water bodies and for groundwater pollution. Decreasing agricultural activities during the last 3 years have caused a slight improvement in the water quality of inland waters and aquifers.

(7) A well-developed network of ecological compensation areas has been developed, consisting of nature protection areas, forests, bogs, meadows and coastal waters. This network, set in a country of relatively low population density, has played a major part in the maintenance of biodiversity. Nevertheless, the extinction of wooded meadows and alvar meadows could be accompanied by a significant loss of species. It is important that the ecological network be maintained and enhanced during the process of privatization.

(8) The next important step will be to connect the network of ecological compensation areas in Estonia with the European Ecological Network (EECONET) system. Therefore, an optimization of the ecological network at different hierarchical levels, and using a variety of criteria, is needed.

ACKNOWLEDGEMENTS

This study was supported by the Estonian Science Foundation grants No. 692 (1994–1995) and No. 187 (1993–1995). A part of the data originates from studies financed by the International Science Foundation grants No. LCU 000 (1994) and No. LLL 100 (1995).

REFERENCES

Agriculture (1995). *Tallinn*. Statistical Office of Estonia, pp. 129
APHA (1981). *Standard Method for the Examination of Water and Waste Water*, 15th edition. American Public Health Organization, Washington, pp. 1134
Arold, I. (1993). *Estonian Landscapes*. Tartu, University of Tartu, pp. 95
Baldock, D., Beaufoy, G., Bennet, G. and Clark, J. (1993). *Nature Conservation and New Directions in the Common Agricultural Policy*. Report for the Ministry of Agriculture, Nature Management and Fisheries, The Netherlands. Institute for European Environmental Policy, HPC, Arnhem, pp. 224

Bastian, O. (1994). Ansätze der Landschaftsbewertung. Diversität. In Bastian, O. and Schreiber, K.-F. (eds). *Analyse und ökologische Bewertung der Landschaft.* Gustav Fischer Verlag, Jena, Stuttgart, 282–5

Bennett, G. (ed.) (1994). *Conserving Europe's Natural Heritage. Towards a European Ecological Network.* Proceedings of the International conference held in Maastricht, 9–12 November 1993. Graham & Trotman/Martinus Nijhoff, London, pp. 334

Bischoff, N. T. and Jongman, R. H. G. (1993). *Development of Rural Areas in Europe. The Claim for Nature.* Netherlands Scientific Council for Government Policy. Sdu uitgeverij, Plantijnstraat, The Hague, pp 206

Cadastre Book of Estonia and Livonia (1918). (In Russian)

Data of the Estonian Land Cadastre (1990). Tallinn. (In Estonian)

Eesti A + O (1993). *Eesti Entsüklopeediakirjastus.* Tallinn

Estonian Land Use Cadastre (1939). Tallinn. (In Estonian)

Juske, A., Sepp, M. and Shiver, Ü. (1991). Land amelioration activities between the two World Wars. In *Land Amelioration.* Estonian Ministry of Agriculture Information Centre, Tallinn (in Estonian)

Karma, O. (1959). *Outlines of the Development of Land Amelioration in Estonia until 1917.* Tallinn

Kasepalu, A. (1991). *That Which the Master Leaves, the Woods Will Take. Land use in the Estonian village.* Institute of Economics, Estonian Academy of Sciences, Tallinn (in Estonian with English summary)

Külvik, M. (compiler) (1993). Environmental status report 1993: Estonia. In *Environmental Status Reports: 1993, Vol. 5: Estonia, Latvia, Lithuania.* IUCN East European Programme, Cambridge, pp. 1–77

Land Balance of the Estonian SSR 1966 (1967). Estonian Board of Statistics, Tallinn. (In Russian)

Land Cadastre of the Estonian SSR 1945–1985 (1986). Estonian Board of Statistics, Tallinn. (In Estonian)

Lippmaa, T. (1935). Main Features of Estonian Geobotany. *Acta et commentationes Universitatis Tartuensis* A XXVII, 4: 1–151. (In Estonian with summary in French)

Mander, Ü., Kuusemets, V. and Ivask, M. (1995a). Nutrient dynamics of riparian ecotones: A case study from the Porijõgi River catchment, Estonia. *Landscape and Urban Planning,* **31**: 333–48

Mander, Ü., Kuusemets, V. and Treier, K. (1994a). Variation in groundwater quality in South Estonian rural areas. In *Groundwater Quality Management* (Proceedings of the GQM 93 Conference held at Tallinn, Sept. 1993). IAHS Publ. No. 220, 81–93

Mander, Ü., Metsur, M. and Külvik, M. (1989). Störungen des Stoffkreislaufs, des Energieflusses und des Bios als Kriterien für die Bestimmung der Belastung der Landschaft. *Petermanns Geographische Mitteilungen,* Gotha, **4**: 233–44

Mander, Ü. and Palang, H. (1994). Changes of landscape structure in Estonia during the Soviet period. *GeoJournal,* **33**(1): 45–54

Mander, Ü., Palang, H. and Jagomägi, J. (1995b). Ecological networks in Estonia. Impact of landscape change. *Landschap,* **3**: 27–38

Mander, Ü., Palang, H. and Tammiksaar, E. (1994b). Landscape changes in Estonia during the 20th century. In Analysis of Landscape Dynamics – Driving Factors Related to Different Scales. EUROMAB Research Program *Land-use Changes in Europe and their Impact on Environment.* Symposium, Bad Lauchstädt near Halle/Saale, Germany, Sept. 27–Oct. 1, 1993. UFZ-Umweltforschungszentrum, Leipzig–Halle, pp. 73–97

Metsur, M. (1993). *Agricultural Non-Point Pollution with Nitrogen.* Thesis for MSc degree in Geology. Tartu, pp. 57 (Manuscript in Estonian with English summary at the Institute of Geology, University of Tartu)

Mikk, M. and Mander, Ü. (1995). Species diversity of forest islands in agricultural landscapes of southern Finland, Estonia and Lithuania. *Landscape and Urban Planning*, **31**: 153–69

Öispuu, S. (compiler) (1992). *Eesti ajalugu Šrkamisajast tŠnapŠevani*. Tallinn, Koolibri

Palang, H. (1994). *The Dynamics of Landscape Diversity and Land Use Structure in Estonia during the 20th Century.* Thesis for MSc degree in Geography. Tartu, pp. 60 (Manuscript in Estonian with English summary at the Institute of Geography, University of Tartu)

Peterson, K. (compiler) (1994). *Nature Conservation in Estonia*. Tallinn, Huma, pp. 48

Pielou, E. C. (1966). The Measurement of Diversity in Different Types of Biological Collections. *J. Theoret. Biol.,* **13**: 131–44

Rodoman, B. B. (1974). Polarization of landscape as a manage agent in protection of biosphere and recreational resources. *Resources, Environment, Settlement.* Moscow, Nauka, 150–62 (In Russian)

Sepp, K., Hansen, V. and Mander, Ü. (1994). Environmental legislation and questions of property rights in Estonia. In *Proceedings of the International seminar on Environment and Properties Rights*, Ligatne, Latvia, 12–16 December 1994, pp. 1–12

Varep, E. (1964). The landscape regions of Estonia. Publications on Geography IV. *Acta et commentationes Universitatis Tartuensis*, **156**: 3–28

Yearbook of the Estonian Land Cadastre 1992 (1993). Tallinn. (In Estonian)

CHAPTER 10

LAND-USE CHANGE IN THE AGRICULTURAL REGION OF WIELKOPOLSKA, POLAND

L. Ryszkowski and S. Bałazy

INTRODUCTION

The purpose of this chapter is to show changes in the land-use pattern in the Wielkopolska region over time. We indicate how these changes influence fluxes of matter and energy in the landscape and discuss measures of control for the negative effects of land-use changes.

The physiographically defined region of Wielkopolska (Great Poland) is located in the basin of the middle part of the Warta river with its tributaries – the Prosna, the Obra and the Noteć. Wielkopolska has an area of approximately 35,000 sq. kilometres. Almost 90 percent of its area is between 50 and 150 m above sea level. The surface of the region was shaped by the Pleistocene glaciations, the last of which regressed some 11 thousand years ago. In Wielkopolska, lowlands with slightly modified ground moraines, together with drainage valleys and hilly or undulating uplands can be distinguished. Although the differences in altitude are not significant, most of the land of Wielkopolska is characterized by an abundance of geo-morphological features such as terminal, ground and lateral moraines, kames and eskers, drumlins, proglacial stream valleys, glacial channel systems, sandy outwashes and extensive flat plateaux (Krygowski, 1961). Loose sandy soils, coarse sandy soils, loamy sands and boulder loams are the most common soils. The majority of soils have good infiltration rates and small storage capacities of water.

The climate of the region is shaped by conflicting air masses coming from the Atlantic (polar-maritime air masses) and eastern Europe and Asia (polar-continental), modified by arctic and tropical air jets. The low stability of climatic conditions is characteristic of the region. Western polar-maritime influxes of air masses bringing moisture from the Atlantic Ocean predominate (Woś, 1994). Since the topography of the region does not directly influence the movement of the air masses, the western winds dominate the region. The mean annual wind velocity is about $4 \text{ m} \times \text{s}^{-1}$, which is the speed at which wind erosion processes start. The net radiation value for the region, together with eastern parts of the Middle Poland Lowland, is the highest in the country. The region's mean long-term temperature values for January range from −2.8 °C to −1.5 °C, for July (the warmest month) from 17.6 °C to 18.0 °C, and for the whole year from 7.6 °C to 8.2 °C. The Wielkopolska region has the lowest precipitation in the country, ranging between 500 mm to 600 mm. Fortunately the distribution of precipitation is favourable for agriculture

189

(maximum in summer and minimum in winter), although a drought often occurs in spring (Woś, 1994). Often low winter precipitation does not compensate for water shortages caused usually by higher evapotranspiration than precipitation during the growing season. Generally low soil water retention makes the situation even worse. The shortage of water in the Wielkopolska region is reflected by very low river outputs in relation to precipitation. According to the estimates of Pasławski (1992), only 22.8 percent of the annually received precipitation is discharged by the Warta river. The annual output from five watersheds – in an area of 8920 sq. km or 25 percent of Wielkopolska – is less than 20 percent of the received precipitation.

At present within the limits of the Wielkopolska region there are over 10,000 settlements, nearly 190 of which are towns. These are mainly small towns of up to 10,000 inhabitants. In the eastern part of the region are large, brown coal, open-cast mines. A number of different industries in Wielkopolska, such as pottery, building and glassworks, use local resources. The heavy machine industry plays a very important role in the region, while chemical plants give rise to many environmental problems. The food industry is predominant. The biggest urban–industrial agglomeration of the region is Poznań, followed by four other centres of lesser industrial importance. Today the population of Wielkopolska is about 3.5 million, including the Poznań population of almost 600,000.

Agriculture is a crucial element in the economy of Wielkopolska. Although light soils predominate, with good skills farmers can obtain favourable yields. Some farms, for example, can yield 6 tonnes of grain per hectare or even more.

Despite a long history in the development of settlement, and the extent to which natural ecosystems have been modified, the Wielkopolska region still contains many areas worthy of nature protection. Variations in topography and soil, together with the structure of the existing plant cover, preserve a high biodiversity including a number of protected plant and animal species.

IMPACT OF HUMAN POPULATION ON LIVING RESOURCES AND THE ENVIRONMENT IN THE PAST

The gradual warming of the climate in Wielkopolska after the turn of the eleventh and tenth millennia BC caused a slow expansion of forest into the tundra. These slow changes were accompanied by the withdrawal of reindeer hunters from the north and the appearance of hunter–gatherer communities, as well as fishermen, in the Wielkopolska region (Kołodziejski, 1995). The activities of these people were well-balanced as regards the regenerative capacities of ecosystems, and no impacts on the environment can be detected. Only since the middle of the Neolithic age (about 5000 years BP) can the first traces of human impacts on the forests be noted and these were connected with the development of agriculture (Broda, 1985). Five kinds of wheat, barley, pea and flax were cultivated and cows, sheep, goats and pigs were bred (Żak,

1978). In the early Iron Age (2550–2400 BP) cultivation and stock breeding reached such a level of organization that the people of the Lusatian tribe were able to build a fortified settlement on a 2-ha island surrounded by a 6 m high rampart, including about 102–106 wooden houses intersected by streets. About 1000 people lived in this stronghold called Biskupin. There were other strongholds of this kind in the region, the biggest one probably being inhabited by about 2500 people (Topolski, 1969). Pollen analyses carried out for this period indicate a steady decrease of the forest area.

Archeological studies indicate that the fertile alluvial soils of the river valleys, and small plateaux located among the wetlands, were colonized first for agricultural purposes (Kurnatowski, 1975). This activity led to the conversion of forests into cultivated fields and started the intensification of water erosion and the loss of the fertility of the alluvial soils. Shifting (slash-and-burn) agriculture was replaced by the alternating two-field system during the seventh century (Topolski, 1969). This consisted, in essence, of a drastic shortening of the fallow period and the substitution of natural processes of regenerating soil fertility by inputs of manure from farm animals. Any imbalance in the replenishment of nutrients resulted in a loss of soil fertility. The farmers cultivating the very fertile alluvial soils were only partially successful. Between the sixth and the tenth centuries, about 60 percent of the settlements in the valleys lasted for a period of 100–150 years. When the soil was exhausted they were abandoned and cultivated fields were turned into pastures (Kaniecki, 1991). Field cultivation was moved to the uplands where forests were cut down. The removal of the permanent plant cover enhanced water runoff and associated water erosion problems. These processes altered soil properties, instigated a change to the heat balance structure of the region, began the progressive drying of soils on the plateaux, and caused valley soils to receive more water to the extent that some were turned into marshes. In the Warta river valley, the main river of the region, the height of the flood plain increased by about 3 m between the tenth to the end of the nineteenth century, because of the soil transported by water erosion (Kaniecki, 1991).

In the tenth century the Polish state was formed by the unification of the tribes living in Wielkopolska. The economic growth of the region accelerated with consequences for land-use changes. Not only were many towns, strongholds and settlements built, but the forest area was progressively converted into cultivated fields. By the eleventh century, cultivated land made up approximately 15 percent of the Wielkopolska area; peasants produced about 45,000 tonnes of grain, while the population density was 4–5 persons per sq. kilometre (Topolski, 1969). In the middle of the fourteenth century Wielkopolska produced over 100,000 tonnes of grain annually.

The change from a two-field rotation pattern to a three-field system was an important factor in the increase of agricultural production. The cultivation of over-wintering and spring crops allowed a shortening of the time when fields lay fallow and thus production was intensified. Under the three-field rotation

191

system the problem of replenishing soil nutrients, which were removed with crop harvests, became more important than in the two-field system. As a result of agricultural expansion in the fourteenth century, the proportion of forested area dropped to about 50 percent of the Wielkopolska area, while the population density increased to about 10 persons per sq km. By the end of the sixteenth century the density had risen further to 21 persons per sq km, while global annual grain production amounted to 240,000 tonnes (about 850 kg per ha). In the first half of the sixteenth century the decline of the forest area was again accelerated, on this occasion by the export of timber (Topolski, 1993). At the beginning of the nineteenth century huge forest areas were cut because of Napoleon's campaign (Broda, 1985), so that the forested area of Wielkopolska had declined to 28 percent by 1838 (Błaszyk, 1976). According to Błaszyk (1976), during the period between 1838 to 1938, the number of forest units above 5,000 ha declined by almost four fold (from 23 to 6). At the same time the number of small forests with an area below 10 ha increased from 1622 to 5047. Thus, the process of forest fragmentation is evidenced, with important consequences for the elimination of big predators such as the wolf and lynx.

In the second half of the nineteenth century programmes of afforestation started and forest losses became less intensive. In 1924 the area covered by forests amounted to 19 percent of the total area and at present the figure is 22 percent.

Accelerated wind and water erosion are two of the side effects of deforestation, leading to soil degradation. Since erosion is a natural process controlled by plant cover, any removal of permanent vegetation in association with unwise farming practices can lead to the increased removal of soil which has accumulated over many millennia. Analysis of the composition of bog sediments clearly indicates a marked increase in erosion rates in the central part of the Wielkopolskia (Osieczna) over the last thousand years (Borówka, 1990). These data match the historical evidence of a decrease in the forest cover. The denudation rate was equal to 0.3 mm per 100 years at the beginning of the eleventh century, steadily increasing up to the fourteenth century when its rate was 1.7 mm per 100 years. After that time there were periods of decreasing and increasing denudation rates, with the highest rate at 4.7 mm per 100 years in the twentieth century when agricultural activities became very intensive (Borówka, 1990).

Despite the flatness of Wielkopolska, and climatic conditions which do not lend themselves to intensive erosion processes, changes in soil profiles caused by denudation processes over a long period can be recognized. Marcinek (1994) found that in Wielkopolska about 673,300 ha of soil show slightly or moderately eroded soil profiles, while 215,800 ha indicate very marked degradation. The decline of soil fertility due to erosion can be compensated by the input of fertilizers, but the loss of soil depth for plant growth cannot be compensated by modern agricultural practices. Thus, measures against erosion are needed for the sustainable development of rural areas, one measure being the modification of the open agricultural landscape by planting mid-field rows of trees or shelterbelts. In Poland, General Dezydery Chłapowski was the first to introduce the idea of differentiation of the agricultural landscape by planting shelterbelts. In the first half of the previous century he established a network of shelterbelts on about 10,000 ha in Turew, central Wielkopolska, in order to control wind erosion. This modification of the landscape to control erosion, modify microclimatic conditions, as well as save moisture in adjoining fields, has been undertaken in other places in Wielkopolska since Chłapowski's activities. At present, a planting programme is supported by district authorities and is developing well.

LAND-USE CHANGES AND LAND DRAINAGE IN WIELKOPOLSKA

Growth in population brought about the colonization of the wetlands for agricultural purposes and attempts were undertaken to drain about 20 percent of the Wielkopolska area (Kaniecki, 1991). This activity, slow at the beginning (in the seventeenth and eighteenth centuries), accelerated towards the end of the eighteenth and the nineteenth centuries, when land drainage was conducted on a large scale. In the same period, engineering work on the regulation of the Warta and Obra rivers started. Before regulation, the river bed of the Warta had a slope of less than 0.2 percent; the water current was slow and the river had many meanders (Kaniecki, 1991). The straightening of river courses, deepening of riverbeds and raising of embankments all facilitated drainage of water from the region. Thus, by the beginning of the present century, the hydrology of the region had been effected by deforestation, drainage of wetlands and regulation of the courses of the main rivers. As shown by Kaniecki (1991), these activities resulted in a decrease in the ground water level by 105 cm in the vicinity of Poznań. During summer the groundwater level falls by more than 2 m in the river valleys. The persistence of low water in the Warta river has increased from 75 days to 102 days per year. Some lakes have disappeared and in others the water level has decreased by 13 percent on average. During a period of 50 years (from 1890 to 1941), 56 percent of the 11,068 small water bodies vanished. Many peat soils have lost substantial amounts of organic material and nowadays they have become infertile marginal lands (Kaniecki, 1991).

The process of land drainage is also reflected in changes to plant communities: a larger number of xerothermic plant species has become evident. For example, Czubiński (1956) found that in the Wielkopolska region xerothermic plant species make up 14 percent of the total number of vascular plants, whereas in the Mazurian region, located to the east, xerothermic species make up less than 7 percent of the total flora.

Denisiuk *et al.* (1992) have analysed changes in the distribution of wet (flooded for a long time), humid (periodically flooded) and dry (never flooded) grasslands. They found a dramatic conversion of wet to dry grasslands: after a period of 20 years (from 1970 to 1989) only 30.6 percent of grassland was left, having previously covered 69.3 percent of the area. At the same time the contribution of dry grassland to the total area of grassland increased from 12.2 to 47.6 percent.

According to Grynia (1962), the drainage of marshlands along the river valleys has led to losses of phosphorus and potassium from soils, which in turn has contributed to the transformation of Wielkopolska plant communities into *Molinia* grasslands. The *Molinia* grasslands show low productivity, providing only one annual hay harvest of poor quality.

Another phenomenon connected with the progressive decrease of the ground-water table in the Wielkopolska region is the transformation of grassland into arable fields. During the first half of twentieth century, for instance, about 20 percent of the grassland in the Prosna river valley was ploughed, and the hay yield in the remaining grassland dropped from 4.0 tonnes to 1.2 tonnes per hectare (Grynia, 1962). Thus, drainage of wetlands, together with regulation of the rivers, has resulted in a serious environmental threat to the region, including the loss of many productive grassland habitats, soil degradation processes, and loss of water resources.

INTENSIFICATION OF AGRICULTURAL PRODUCTION

The deforestation and drainage of Wielkopolska are not the only develop-ments which have had important environmental consequences. Changes in farming practices that accompanied modernization of agriculture resulted in a substantial increase of production but also created new environmental problems.

In the nineteenth century, as explained, intensification of agricultural production was achieved by elimination of fallow within the alternating three-field system. Crops were grown for green manure, such as fodder crop mixtures of alfalfa, clover, serradella and lupin which regenerated soil fertility and simultaneously provided feed for animal husbandry. The old three-field system was replaced by a crop rotation system lacking fallow. Also in the nineteenth century, animal breeding was quite environmentally balanced and provided farm manure to maintain soil fertility. The cultivation of potatoes was widely expanded and, to some extent, compensated for poor yields of cereals when climatic conditions were unfavourable. But soil erosion processes were intensified (Szumański, 1977). On the other hand, expanded cultivation of potatoes provided more feed for the increased husbandry of animals, which in turn resulted in higher supplies of farmyard manure. Thus in regions having light soils, potato cultivation played an important role in stimulating the prosperity of farming (Pietruszyński, 1937). Farmyard manure, green manure, as well as diversified crop rotation patterns, not only stimulated higher yields but also improved soil quality. Nevertheless, diversified crop rotations were only introduced slowly. In Wielkopolska, the proportion of fallow on arable land decreased from 20 percent to about 10 percent within the period from the end of the nineteenth century to 1929. In 1938 the fallow land still covered 6.8 percent of the total arable area (Baranowski *et al.*, 1970).

A slow increase in production can be observed in the first decades of the twentieth century, reaching per hectare yields of 1.45 tonnes for rye, 1.76 tonnes for wheat, 1.75 tonnes for barley, 14.2 tonnes for potatoes, 21.0 tonnes for sugar beet and 1.2 tonnes for rape seed in 1931–1934 (Pietruszyński, 1937). The land reform carried out in 1945–46 resulted in a farm structure of individual holdings: the percentage of the agricultural area owned by private farms ranged from 50 to 80 percent of the total agricultural area in different parts of Wielkopolska (Tyszkiewicz, 1978). By 1970, 50 to 80 percent of the individual farms were larger than 10 ha, and by 1984 the proportion of land owned by private farmers amounted to between 64 and 90 percent in the different regions.

The increased production from individual farms was achieved by intensification of labour, while on state farms an intensification in production methods was the driving force for increase in yields (Stola, 1978). The use of fertilizers increased at a much higher rate, and by the 1970s reached a level that was about two times higher on state farms compared with individual holdings (Stola, 1978). The highest applications of mineral fertilizers, reaching 240 kg per ha of N, P_2O_5 and K_2O, were achieved in 1987 and 1988. After the economic crisis of the 1980s, and the change of political system in 1989, the application of fertilizers rapidly decreased, reaching less than 100 kg per ha of N, P_2O_5 and K_2O at the beginning of the 1990s. A slight increase has been evident in the last two years. In 1991 the yields of cereals varied from 3.4 tonnes per ha to 4.27 tonnes per ha from area to area, those of potatoes from 16.3 to 17.6 tonnes per ha, and those of sugar beet from 30.6 to 35.0 tonnes per ha (Agricultural Statistical Yearbook, 1993). Thus, following World War II, the agriculture of Wielkopolska can be characterized as one of the most progressive in the whole of Poland.

Intensive inputs of artificial fertilizers, together with large quantities of liquid manure produced by industrialized animal husbandry have caused threats to the environment and particularly a deterioration in water quality. Non-point sources of pollution caused by extensive leaching of high quantities of artificial fertilizer and liquid manures, as well as other chemicals, into the groundwater have recently been recognized as one of the most important factors reducing the quality of inland waters (Ryszkowski, 1992). The following three examples illustrate the intensity of this threat to the environment of the Wielkopolska region.

Bartoszewicz (1990) found an increase of nitrates in the water of drainage canals in the agricultural landscape of the central part of Wielkopolska (Turew) from 1972 to 1990. Also, there was clear evidence of an increase of phosphates and potassium in the water in the 1980s compared with the 1970s

Table 10.1   Mean concentration of elements (mg dm⁻³) in drainage canals located in
an agricultural landscape – Turew, Poland

Period	*1972–1976*	*1982–1985*	*1986–1990*
Number of samples	295	332	202
Elements			
$N\text{-}NO_3^-$	12.60	18.60	48.30
$P\text{-}PO_4^{3-}$	0.18	0.37	0.31
K^+	3.20	7.80	6.10

After Bartoszewicz (1990)

(Table 10.1). In the 1980s fertilizers were applied in this region in doses of about 300 kg per ha of N, P_2O_5 and K_2O, and heavy applications of this type resulted in increased leaching of nutrients into the drainage canals.

Zerbe *et al.* (1994) studied water chemistry of 125 wells distributed over Poznań province; they found that only 20.8 percent were not polluted, taking the accepted standard of $N\text{-}NO_3^-$ as 10 mg dm⁻³. The situation for potassium was even worse, while the mean concentration of $N\text{-}NO_3^-$ increased from 12.9 mg dm⁻³ in the 1984–1986 period to 35.8 mg dm⁻³ in 1992 (Zerbe *et al.*, 1994). These results correspond very well with the estimates of Bartoszewicz (1990).

Inspection of water from wells on farms in the Wielkopolska region shows that, during the last five years, water has been of poor quality. In 1993, for example, water of poor quality was found in 62 to 81 percent of the sampled wells (Ochrona Środowiska, 1994). The reported results of water quality analyses clearly indicate that groundwaters are polluted and that diffuse pollution is an increasing threat to the environment in the region.

After the change in political system, state farms underwent a process of privatization. The Polish Parliament entrusted the process of restructuring and privatization of state agriculture to the State Treasury Agricultural Property Authority (STAPA); this agency started active work in the middle of 1992. Up to the end of 1995, 321,778 ha of state farms in the Wielkopolska region had been taken over by the STAPA: 22.5 percent of which was sold, 57.7 percent taken on lease, and 22.5 percent placed under the administration of the STAPA as cultivated fields. Only 1.1 percent (3,618 ha) of the cultivated land taken over by the STAPA has been converted into afforestation schemes or given national park status. Consequently, recent changes in the ownership of cultivated land has had little impact on the basic land-use structure of the Wielkopolska region.

CONTROL OF THE NEGATIVE SIDE EFFECTS OF LAND-USE CHANGES

Land-use changes in Wielkopolska, brought about by expansion of human settlement, intensification of agriculture, land drainage and fragmentation of the forest area have all contributed to accelerated rates of water and wind erosion and a lowering of groundwater levels, with consequences for changes in plant communities, especially grasslands. The intensification of agricultural production during the second half of the twentieth century has also enhanced problems of diffuse pollution which, in combination with urban and industrial pollution, has caused a deterioration in water quality of the region.

One of the great challenges of the present day is to restrain the trend towards environmental degradation. By taking advantage of available knowledge on the functioning of ecosystems, it is possible to formulate proposals on ecological guidelines for landscape management aimed at: restoring the cycling of water resources; controlling water pollution; and counteracting impoverishment of natural resources caused by the intensification of agricultural production. These three directions in landscape management will be discussed on the basis of results of investigations carried out in the Research Centre for Agricultural and Forest Environment in Poznań (Poland).

Influence of plant cover structure on water cycling

The heat balance of ecosystems, watersheds or any other large region (denoting the partitioning of solar energy for various work performed in the system under consideration – e.g. energy used for evapotranspiration or for the heating of air or soil) is the driving force in the water balance of an area. It must be borne in mind that any change in heat balance inevitably brings about changes in water balance and this should be of paramount importance for management of water resources in a region. By inducing structural changes in plant cover of an agricultural landscape, it is possible to change the heat balance and thereby influence the intensity of evapotranspiration. For example, by planting shelterbelts in a landscape consisting solely of cultivated fields, not only is the heat balance modified (influencing the water cycle), but also, by changing the wind speed in the layers of air close to the ground, significant changes in potential evapotranspiration are created. The real evapotranspiration of the area intersected by shelterbelts is affected. Indeed, introduction of shelterbelts among cultivated fields in experimental regions of southern Wielkopolska resulted in a decrease in potential evapotranspiration from adjoining cultivated fields. This amounted to 70 mm on average during one period, while the real evapotranspiration of the total landscape increased by 20 mm (Ryszkowski and Karg, 1976; Ryszkowski and Kędziora, 1987). The established rise in real evaporation rates was due to much higher evapotranspiration from trees in the shelterbelts in comparison to evapotranspiration from areas lacking shelter-belts. Hence, planting of shelterbelts produces a double effect: it not only intensifies the water cycle in the landscape but also facilitates a considerable preservation of water resources in the sheltered part of the agricultural area.

The complicated interaction between energy fluxes and the water cycle can be illustrated from the studies of Kedziora *et al.* (1989). The planting of shelterbelts on 12 percent of the total area of a studied watershed led to a decrease in energy used for evapotranspiration from the area between shelterbelts. Annual evapotranspiration in those areas decreased from 510 mm to 460 mm.

The importance of shelterbelts for water management is particularly evident during hot and windy weather, where advective heat transport plays an important role. During such hot and windy periods the water savings in regions rich in shelterbelts can amount to as much as 40 mm, and the potential evapotranspiration might decrease by 34 percent compared with unsheltered fields (Ryszkowski and Kędziora, 1987).

Analyzing the partitioning of energy for various kinds of work performed within the ecosystem (heat balance), enormous differences among various ecosystems in the landscape can immediately be detected (Table 10.2). In the studied landscape the values of energy used for evapotranspiration differ by 75 percent (shelterbelt compared with bare soil), while the difference in energy used for heating air comes to 438 percent (shelterbelts compared to bare soil). The values of energy used for heating soil differ by 200 percent (shelterbelt and meadow). Disregarding bare soil, because of its ecologically artificial condition, one can still observe a high diversity between ecosystems. Thus, the shelterbelt uses about 39 percent more energy for evapotranspiration than wheat fields, while wheat fields divert about 218 percent more energy to heat air than the shelterbelt ecosystem (Table 10.2).

The magnitude of differences caused by the various types of plant cover becomes evident by comparing shelterbelts and meadow ecosystems. For evapotranspiration, shelterbelts use 122 percent of the energy which is used for comparable processes in the meadow ecosystem. For heating soil 300 percent, and for heating air only 56 percent, of the energy is used compared to the meadow ecosystem (Table 10.2). Thus, shelterbelts influence evapo-

Table 10.2 Heat balance structure (MJm^{-2}) and evapotranspiration (mm) during the growing season in an agricultural landscape

Parameter	Landscape elements					
	Shelterbelt	Meadow	Rapeseed field	Beet field	Wheat field	Bare soil
Rn	1730	1494	1551	1536	1536	1575
LE	1522	1250	1163	1136	1090	866
A	121	215	327	339	385	651
S	87	29	61	61	61	47
LE:Rn	0.88	0.84	0.75	0.74	0.71	0.55
E	609	500	465	454	436	346

[a]Rn = net available energy; LE = latent heat flux; A = air-sensitive heat flux; S = soil-sensitive heat flux; E = evapotranspiration in mm. Modified after Ryszkowski and Kędziora (1987)

transpiration much more than grassland and simultaneously exert a cooling effect on the air; but they heat the soil to a greater extent than grassland.

The restructuring of a landscape of uniform fields into a mosaic of land covers will cause changes in local heat balances of the ecosystems forming the more diversified landscape; this process in turn will effect local movement of air masses and intensity of water cycles. The results of ecological investigations provide a basis for elaboration of ecological guidelines for management of the water cycle in agricultural landscapes by differentiation of the structure of its plant cover.

Change in drainage practices in Wielkopolska could be developed simultaneously with a programme of restructuring the agricultural landscape by planting networks of trees. The present drainage system of cultivated fields is aimed at removal of water from the landscape, especially after early spring thaws and rains. Tile drains carry water into drainage channels from where the water flows to rivers and the sea. A water storage system based on huge reservoirs is not sufficient to maintain water reserves for the warmer period of the plant growth season. That is why the guidelines developed on an eco-policy for rural areas (Michna and Ryszkowski, 1995) recommended storage of water in small field reservoirs made in depressions in the landscape, instead of losing water from the landscape. Drainage pipes could be emptied into small reservoirs without outlets in order to store water for the drier season. Nevertheless, small field reservoirs should have devices permitting release of water into channels during very rainy years.

Control of groundwater pollution

Recently, it has been shown that shelterbelts (mid-field rows of trees or small forested blocks) and strips of grassland help to control transportation of various chemical compounds from cultivated fields into water basins (Peterjohn and Correl, 1984; Pauliukevicius, 1981; Ryszkowski and Bartoszewicz, 1989; Ryszkowski, *et al.*, 1989). Shelterbelts and grassland strips thus function as biogeochemical barriers controlling the spatial spread of pollutants. The information summarized by Bartoszewicz (1990) and Ryszkowski and Bartoszewicz (1989) indicates that shelterbelts and grassland can control groundwater transportation of many inorganic ions such as $N-NO_3^-$, Cl^-, SO_4^{3-}, PO_4^{3-}, K^+, Na^+, Ca^{2+}, Mg^{2+} and others. In all of these studies, watershed fragments were selected so that shelterbelts and grassland were situated in places where groundwater outflow from neighbouring cultivated fields passed under them. In all cases groundwater was within direct of indirect reach of plant root systems.

The efficiency of control of chemical compounds by various biogeo-chemical barriers varies considerably (Table 10.3). Nevertheless, in all cases studied by Bartoszewicz (1990) a decline in phosphate concentration was observed when groundwater passed under biogeochemical barriers. Ryszkowski

Table 10.3 Impact of biogeochemical barriers on P-phosphate concentration (mg dm^{-3}) in groundwater

| | | | Mean phosphate concentration in groundwater | | |
Ecosystem	Period	Sample number	Field (F)	Biogeochemical barrier (B)	F – B
field–meadow I	1986–1989	47	0.050 ± 0.016	0.030 ± 0.012	0.020
field–meadow II	1987–1989	26	0.252 ± 0.044	0.051 ± 0.034	0.201*
field–meadow III	1987–1989	31	0.079 ± 0.022	0.037 ± 0.014	0.042*
field–shelterbelt I	1984–1986	26	0.190 ± 0.048	0.081 ± 0.022	0.109*
field–shelterbelt II	1986–1989	34	0.064 ± 0.020	0.049 ± 0.023	0.015

*difference statistically significant $p < 0.05$; ± standard deviation
After Bartoszewicz (1990)

et al. (1989) show that in cases where cultivated fields directly adjoin the drainage channel, annual leaching of phosphate ions amounted to 33–90 mg P m^{-2}. When cultivated fields were separated from the drainage channel by an 80 to 90 m grassland strip, annual leaching of phosphates amounted to only 12–37 mg P m^{-2}. Thus, output of chemical compounds of the watershed depends on plant cover structure. Taking into account all available information on the functions of biogeochemical barriers, one can conclude that shelterbelts and grassland are an effective means of control of ground water pollution.

Influence of landscape structure on biodiversity

Recent results (Ryszkowski and Karg, 1991; Ryszkowski *et al.*, 1993) show that a mosaic landscape, composed of small cultivated fields, shelterbelts, stretches of grassland, small ponds and other non-agricultural landscape components, supports a higher richness and diversity of insect communities compared with a uniform landscape, composed of large cultivated fields nearly free from non-agricultural landscape components. Analyses indicate that a more diversified landscape structure stimulates a higher number of taxa, as well as increasing the density and biomass of insect communities. There is also a richer bird community in the mosaic landscape compared (Gromadzki, 1970). Further, these studies indicate that, at least in some animal communities, impoverishment of fauna due to intensive farming can be mitigated by altering the structure of the landscape by introducing a network of shelterbelts, stretches of grassland, small ponds and other refuges. These results have an important bearing on the protection of living resources in the rural landscape.

CONCLUSION

The results of the studies discussed above indicate that introduction of shelterbelts, stretches of grassland and other natural elements increasing the heterogeneity of landscape enhance water storage, control ground water chemistry and counteract impoverishment of biota due to intensification of agricultural production.

The tendency towards a rapid increase in agricultural production very often causes an undue exploitation of natural resources, which in turn creates environmental problems. The increasing recognition of the natural laws of ecosystem functions facilitates a more objective evaluation of alternative technologies which, at the same time, optimize agricultural production, nature and environmental protection and meet social needs. This is the programme for the sustainable development of the countryside. An agroecosystem approach is useful in developing a better understanding and control of environmental threats to rural areas. Thus a programme of nature preservation and environmental protection should involve introduction and use of biogeochemical barriers as a supplement to the technical measures of pollution control. Ecological agriculture does not lead to a refutation of modern means of agricultural production, but to a more rational agroecosystem strategy which seeks to meet jointly the objectives of food production and nature conservation. Comprehensive management of ecological processes, together with technical measures aimed at control of environmental threats to the countryside, not only enhance the success of environmental protection but also increase the effectiveness of nature conservation.

In this paper the following environmental threats have been recognized as most important for the Wielkopolska region:

(1) Increasing water deficits that manifest themselves in falling ground-water tables, disappearing natural water reservoirs and diminishing wetland areas.

(2) Low water quality caused by urban and industrial contamination, supplemented by increasing non-point sources of pollution caused mainly by intensive agriculture.

(3) Soil degradation primarily as a result of the loss of organic matter.

(4) Impoverishment of biological resources caused by simplification of agricultural landscapes and application of intensive agrotechnologies.

The following activities have been proposed as the most important measures to control these threats (Michna and Ryszkowski, 1995; Ryszkowski and Bałazy, 1995):

(1) Increase water storing capacity of the landscape by introduction of mid-field networks of trees, formation of small water reservoirs with controlled outlets, and enhancement of humus resources.

(2) Development of efficient techniques for treatment of animal wastes, combined with measures to control diffuse pollution using

biogeochemical barriers at the watershed level. Technical and ecological systems of pollution control should enhance their efficiencies by reciprocal support.

(3) Increase erosion control by propagating anti-erosion tillage practices and crop rotation patterns, as well as by restructuring the landscape with the help of mid-field tree planting and grassland strips.

(4) Augment application of organic fertilizers and crop rotation patterns to enrich organic matter of soil.

(5) Protect or restore refuge habitats for biota and ensure their connectivity by a network of ecological corridors.

(6) Adjust application of fertilizers and pesticides to levels that balance with soil conditions and plant protection needs.

ACKNOWLEDGEMENT

The management of agricultural landscape according to these principles has been developed under the supervision of the Research Centre for Agricultural and Forest Environment of the Polish Academy of Sciences in the D. Chłapowski Agroecological Landscape Park, on an area of 17,000 ha. The implementation of the described guidelines for sustainable development of agriculture is proposed in the eco-policy for rural areas accepted by the regional authorities.

REFERENCES

Agricultural Statistical Yearbook (1993, 1994). Główny Urzad Statystyczny, Warszawa, pp. 373

Baranowski, B., Dydowiczowa, J., Leskiewicz, J. and Topolski, J. (1970). *Zarys historii gospodarstwa wiejskiego w Polsce*. Vol. 3 Państwowe Wydawnictwo Rolnicze i Leśne, Warszawa, pp. 801

Bartoszewicz, A. (1990). Chemical composition of ground waters in agricultural watersheds under the soil and climatic conditions of the Koscian Plain. (In Polish). In Ryszkowski, L., Marcinek, J. and Kędziora, A. (eds). *Obieg wody i bariery biogeochemiczne w krajobrazie rolniczym*. Wydawnictwo Naukowe Uniwersytetu, A. Mickiewicza, Poznań, pp. 127–42

Błaszyk, H. (1976). The changes of Wielkopolska forest. (In Polish). Roczniki Akademii Rolniczej, Poznań. *Rozprawy naukowe*, **73**: 47

Borówka, R. K. (1990). Denudation process intensity on the Vistulian till plains in relation to prehistoric settlement and human activity. Leszno region, Middle Great Poland. *Quaestiones Geographicae*, **13/14**: 5–18

Broda, J. (1985). The deforestation process on Polish territory since prehistoric times. (In Polish). *Czasopismo Geograficzne*, **42**: 151–71

Czubiński, Z. (1956). Role of xerothermic elements in the plant cover of Wielkopolska. (In Polish). *Zeszyty Problemowe Postepów Nauk Rolniczych*, **7**: 45–50

Denisiuk, Z., Kalemba, A., Zajac, T., Ostrowska, A., Gawliński, S., Sienkiewicz, J. and

Rejman-Czajkowska, M. (1992). *Integration between agriculture and nature conservation in Poland*. IUCN East European Programme. Information Press, Oxford, pp. 162

Gromadzki, M. (1970). Breeding communities of birds in mid-field afforested areas. *Ekologia Polska*, **18**: 307–50

Grynia, M. (1962). *Molinia coerulea* meadows of Wielkopolska. (In Polish). Poznańskie Towarzystwo Przyjaciół Nauk. *Prace Komisji Nauk Rolniczych i Komisji Nauk Leśnych*, **13**: 145–262

Kaniecki, A. (1991). Drainage problems of Wielkopolska Plain during the last 200 years and changes in water relationships. (In Polish). In *Ochrona i racjonalne wykorzystanie zasobów wodnych na obszarach rolniczych w regionie Wielkopolski*. ODR Sielinko, Poznań, pp. 73–80

Kędziora, A., Olejnik, J. and Kapuściński, J. (1989). Impact of landscape structure on heat and water balance. *Ecology International*, **17**: 1–17

Kołodziejski, S. (1995). Primeval history. (In Polish). In Marcinek, J. (ed.) *Kronika dziejów Polski*. Wydawnictwo Ryszard Kluszczyński. Kraków, pp. 7–24

Krygowski, B. (1961). *Geografia fizyczna Niziny Wielkopolskiej*. Część 1. Geomorfologia. (In Polish). Poznańskie Towarzystwo Przyjaciół Nauk, Poznań, pp. 116

Kurnatowski, S. (1975). The early medieval economic break in Wielkopolska and its landscape and demographic consequences. (In Polish). *Archeologia Polska*, **20**: 145–60

Marcinek, J. (1994). Extension of accelerated soil erosion in Great Poland. *Roczniki Akademii Rolniczej w Poznaniu*, **266**: 63–73

Marcinek, J., Komisarek, J. and Spychalski, M. (1990). Soils of Middle Wielkopolska. (In Polish). In Ryszkowski, L., Marcinek, J. and Kędziora, A. (eds). *Obieg wody i bariery biogeochemiczne w krajobrazie rolniczym*. Wydawnictwo Naukowe Uniwersytetu A. Mickiewicza, Poznań, pp. 21–31

Michna, W. and Ryszkowski, L. (1995). Basic activities for the realization of the principles of eco-policy in agriculture. (In Polish). In Ryszkowski, L. and Bałazy, S. (eds). *Zasady ekopolityki w rozwoju obszarów wiejskich*. Zaklad Badań Środowiska Rolniczego i Lesnego PAN. Poznań, pp. 21–8

Ochrona środowiska (1994). *Environment Protection*. Main Statistical Office. Warsaw, pp. 518

Pasławski, Z. (1992). Hydrobiology and water reservoirs in Warta basin. (In Polish). In *Ochrona i racjonalne wykorzystanie zasobów wodnych na terenach rolniczych w regionie Wielkopolski*. ODR Sielinko, Poznań, pp. 5–28

Pauliukevicius, G. (1981). *Ecological role of the forest stands of the lake slopes*. Pergale, Wilno, pp. 191

Peterjohn, W. T. and Correl, D. L. (1984). Nutrient dynamics in agricultural watersheds: observations on the role of a riparian forest. *Ecology*, **65**: 1466–75

Pietruszyński, Z. (1937). Die Pflanzenproduktion in Grosspolen einst und gegenwärtig (In Polish). *Roczniki Nauk Rolniczych i Leśnych*, **42**: 18–370

Ryszkowski, L. (1992). Agriculture and diffuse pollution of the environment. (In Polish). *Postępy Nauk Rolniczych*, **39–40**: 3–14

Ryszkowski, L. and Bałazy, S. (1995). Strategy for environmental and nature protection in rural areas. (In Polish). In Ryszkowski L. and Bałazy, S. (eds). *Zasady ekopolityki w rozwoju obszarów wiejskich*. Zaklad Badań Środowiska Rolniczego i Leśnego PAN. Poznań, pp. 49–64

Ryszkowski, L. and Bartoszewicz, A. (1989). Impact of agricultural landscape structure on cycling of inorganic nutrients. In Clarholm, M. and Bergström, L. (eds). *Ecology of arable land*. Kluwer Academic Publishers. Dordrecht, pp. 241–6

Ryszkowski, L. and Karg, J. (1976). Role of shelterbelts in agricultural landscape. In Missonnier, J. (ed.) *Les bocages – historie, ecologie, economie*. CNRS, Univ. de Rennes. Rennes, pp. 305–9

Ryszkowski, L. and Karg, J. (1991). The effect of the structure of agricultural landscape on the biomass of insects of the above-ground fauna. *Ekologia Polska*, **39**: 171–9

Ryszkowski, L. and Kędziora, A. (1987). Impact of agricultural landscape structure on energy flow and water cycling. *Landscape Ecology*, **1**: 85–94

Ryszkowski, L., Karg, J., Margarit, G., Paoletti, M. G. and Zlotin, R. (1993). Above-ground insect biomass in the agricultural landscapes of Europe. In Bunce, R. G. H., Ryszkowski, L. and Paoletti, M. G. (eds). *Landscape Ecology and agroecosystems*. Lewis Publishers. Boca Raton, pp. 71–82

Ryszkowski, L., Karg, J., Szpakowska, B. and Życzyńska-Bałoniak, I. (1989). Distribution of phosphorus in meadow and cultivated field ecosystems. In Tissen, H. (ed.) *Phosphorus cycles in terrestrial and aquatic ecosystems*. Turner-Warwick Communications, Saskatoon, pp. 178–92

Stola, W. (1978). Labor and capital inputs. (In Polish). In Kostrowicki, J. (ed.) *Przemiany struktury przestrzennej rolnictwa Polski 1950–1970*. Prace Geograficzne 127. Zakład Narodowy Ossolińskich. Wroclaw, pp. 45–109

Szumański, A. (1977). Changes in the courses of the lower San channel river in XIX and XX centuries and their influence on the morphogenesis of its floodplain. *Studia Geomorphologica Carpatho-Balcanica*, **11**: 139–53

Topolski, J. (ed.) (1969). *Dzieje Wielkopolski*. (In Polish with English summary). Vol. 1. Wydawnictwo Poznańskie, Poznań, pp. 1082

Topolski, J. (1993). Natural resources and the history of human beings (In Polish). *Kosmos*, **42**(1): 47–54

Tyszkiewicz, W. (1978). Agrarian structure. (In Polish). In Kostrowicki, J. (ed.) *Przemiany struktury przestrzennej rolnictwa Polski 1950–1970*. Prace Geograficzne 127. Zakład Narodowy Ossolinskich, Wrocław, pp. 15–44

Woś, A. (1994). *Klimat Niziny Wielkopolskiej*. (In Polish). Wydawnictwo Naukowe Uniwersytetu A. Mickiewicza, Poznań, pp. 192

Zerbe, J., Kabacinski, M. and Siepak, J. (1994). Chemical composition of ground waters in the area of Poznań province. (In Polish). *Przeglad Geograficzny*, **42**: 360–4

Żak, J. (1978). Polish land in antiquity. (In Polish). In Topolski, J. (ed.) *Dzieje Polski*. Państwowe Wydawnictwo Naukowe, Warszawa, pp. 14–76

CHAPTER 11

LAND-USE/LAND-COVER CHANGES UNDER AGRICULTURAL IMPACTS AND THE PROSPECTS OF ECOLOGICAL AGRICULTURAL DEVELOPMENT IN THE EUROPEAN PART OF RUSSIA

E. V. Milanova, V. N. Solntsev, P. A. Tcherkashin,
E. Yu. Lioubimtseva and N. N. Kalutskova

INTRODUCTION

The problem of restructuring agriculture in Russia during the period of economic and social reforms is of vital importance. There are two reasons: economic and environmental. The effective use of natural and social resources of Russian agriculture should generate essential economic benefits and save natural resources from exhaustion. However, ecologically sound agriculture should also help to reconcile social needs with requirements of the environment.

Though the importance of the principles and methods of nature management is evident, the relations between nature and society remain unfavourable for both sides:

(1) A sector-based approach to resource management still prevails. Each sector explores nature and makes environmental decisions only within the narrow limits of its economic interest. This results in tremendous distortions between interests of the sector and environmental needs. For example, it is important for a land owner (in the agricultural sector) to increase agricultural production. The fertilization of soils and maintenance of ecological resources are of interest only within the limits of production.

(2) The exploitation of natural resources is based on their economic value without taking into account relations between the different elements of the ecosystem; as a result the system becomes less sustainable.

(3) Production remains the main criterion for our evaluation of the environment. This means that nature management in each country is closely connected with the socioeconomic system and the wealth of its people. The unfavourable economic situation in Russia influences its nature management system and leads to acute environmental problems.

It is important to study natural objects not from the consumers' point of view, but as a whole system, including the elements which do not have a direct economic value but are important for the system's functioning and integrity. A system of land use must be created that does not cause contradictions in existing natural processes, destroy or exhaust the system but which make it more reliable and help in its development.

Here, a complex landscape approach through remote sensing techniques is used to study agricultural impact on landscapes for all of Russia (country scale) and for case studies in the European part of Russia (regional and local scales). Scale-dependent applications are necessary to understand the dynamics of agrolandscapes.

AGRICULTURAL ACTIVITY AND REGIONAL POLICY IN RUSSIA

The agricultural sector in Russia contributes about 16 percent to the country's gross domestic product (GDP) and employs about 13 percent of human labour. Livestock is important in the Russian agricultural economy: approximately 40 percent of the gross value of Russian agricultural output stems from crops and horticulture, with the remaining 60 percent from livestock. Between one-quarter and one-third of the total agricultural labour on collective and state farms is directly employed in the livestock sector.

Russia's total agricultural growth in the 1980s reflects the pattern for the Union as a whole (i.e. a growth of about 2 percent annually from 1980 to 1989, and negative growth thereafter). With approximately 215 million ha, Russia's agricultural area (excluding forests) is larger than Kazakhstan's (198 million ha) and the Ukraine's (41 million ha). Irrigation is, at present, relatively unimportant: about 4 percent of the arable land is equipped for irrigation. The intensity of use of Russian agricultural land is at an intermediate level between the extensive wheat growing and grazing areas of Kazakhstan and intensively cultivated Ukrainian lowlands.

Approximately 53 percent of the cultivated area in Russia is used for growing grain crops. Higher-yielding winter crops are sown on about 30 percent of the grain area, with spring grain on the remaining 70 percent. Other important field crops are: sugar beet, flax, sunflowers, potatoes, vegetables, feed roots, hay and fruit. Russian wheat growing conditions are similar to those in Canada, the United States and Argentina, but yields lag behind those in Canada by about 10 percent. The yields of potatoes, sugar beet and feed, including grasslands and hay, lag behind North American levels much more than grain. Significant increases in yields as well as efficiency gains could be achieved by better management and improved technology. Yield increases in grains would be more modest, but increased efficiency in input use and reduced post-harvest losses could bring significant economic gains.

Yields and efficiency in the livestock sector lag behind the world level more than in the crop sector. Annual milk yields of about 2800 kg per cow reach only 50–60 percent of West European levels and 40–45 percent of US levels. Feeding efficiency in meat production is estimated to be approximately half that of Western Europe. Low yields in the livestock sector are due to a great extent to chronic feed shortages: more animals are kept than can be efficiently fed. The low quality of the feed used and the poor genetic stock of the animals also reduces productivity in the livestock sector.

Russia was the largest food importer of the former Soviet states and it has traditionally been a net importer of most food products, with the exception of eggs, bread products, potatoes and fish. In 1990, prior to the recent fall in demand, Russia imported 13 percent of the meat and 17 percent of the milk it consumed. Between 68 million and 75 million tons of grain have been used for feed in Russia recently, compared with imports of between 18 and 23 million tons (including inter-republic trade). A one-third reduction in demand for grain for feed purposes would thus turn Russia from a net importer into a net exporter of grain.

Food consumption in Russia, on average, remains adequate in terms of calories and average nutrient level, even with the reductions of 1991 and 1992. Regional variations in food consumption were significant under the old distribution system, and it became even more so with the disruption of internal trade at the end of 1991 and the slow response to formal liberalisation of internal trade since 1992.

AGRICULTURAL AREAS AS SOCIO-NATURAL SYSTEMS: A LANDSCAPE APPROACH FOR THEIR STUDY AND MAPPING

Agriculture is a sphere of active interference by society in nature during which natural landscapes are transformed into agrolandscapes of different kinds. Usually agrolandscapes are less stable than natural landscapes, since the stability of any system strongly depends on the diversity of its components. For example, crop-based landscapes are less sustainable compared with livestock-based landscapes: arable land tends to have only one type of vegetation cover instead of the mixed fodder plants associated with livestock systems. Thus the main challenge for agriculture is to combine agro-(transformed) landscapes with natural landscapes.

More than ever, there is a need for rational planning of land use/cover development to make optimal use of land resources (Stomph and Fresco, 1991). But with increasing pressure on the environment caused by agricultural production, a new and complex approach is required to study the 'society–nature' system. The present-day landscapes (PDL) paradigm is one possible approach (Milanova *et al.*, 1993). This landscape approach to the environment, based on a combination of hierarchically nested geosystems (present-day landscapes – PDLs), provides a method for understanding land-use/cover structure and change.

PDLs are defined as units of land surface characterised by a structurally organised combination of natural and economic components, whose close interactions provide a spatially distinct and temporally stable territorial system. This notion is very close to the definition of 'land unit' given by the FAO (1976) in the framework for their evaluation of agricultural activities:

An area of the Earth's surface, the characteristics of which embrace all reasonably stable or predictably cyclic attributes of the biosphere vertically above and below the area, including those of the atmosphere, the soil and the underlying geography, the hydrology, the plant and animal population and the results of past and present human activity to the extent that these attributes exert a significant influence on present and future uses of the land by man.

The landscape approach provides a base for perception of the world as a system of interrelated territorial units with different environmental situations. In response to this issue, a hierarchical landscape classification scheme is proposed for scale-dependent landscape applications. The main points of landscape monitoring and assessment are to make an inventory and diagnosis of their status, based on scale-dependent mapping and geographic information systems (GIS).

Recent extensive scientific work in the field of landscape ecology (Forman and Godron, 1986) has resulted in a number of fundamental concepts for landscapes, or land units, that have proved important for diverse environmental assessment applications. The 'land(scape) unit' is described as an ecologically homogeneous tract of land at the scale at issue. It provides a basis for studying landscape ecological relationships from a topological as well as from a chronological point of view. There are two meanings of the word *landscape*: (1) as a *common language meaning* for a physiognomic category concerned with visual landscape; and (2) as a *scientific language meaning,* a rather specific one, used by geographers and landscape ecologists as an equivalent to a geoecosystem or territorial category (territorial complex).

The anthropogenic alteration of natural land-cover patterns is quite well known – the natural land cover is removed by such activities as agricultural practices, urban development and forestry, and is replaced by managed systems of altered land-cover structures. At present, very few territories of the world can be classified as natural landscapes. Most of them are influenced by human-induced phenomena, such as acid rain and global climatic changes. The resulting landscape is generally a mixture of natural and human-managed patches of different sizes and shapes.

The term 'landscape' is commonly understood within the geographic and landscape ecology research community as designating a territory that integrates different components of the geobiosphere and the socioeconomic sphere. Any landscape system of any spatial dimension currently represents a complex ecological–economic system, where two subsystems – natural and anthropogenic – coexist and interact within the boundaries of their comparatively stable natural bases. Thus the present-day peculiarity of any landscape system is expressed in the character of the anthropogenic transformation of its natural pattern. The relationships between natural and socioeconomic subsystems within present-day landscapes are shown on Figure 11.1.

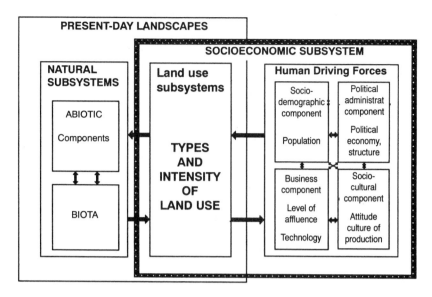

Figure 11.1 Natural and socioeconomic subsystems of present-day landscapes

Landscape analysis represents a holistic view of man-made and natural surroundings and attempts to bridge the gap between natural and human subsystems. Within the framework of present-day landscape mapping it will be possible and relevant to investigate how regulatory tools addressing different parts of the systems will influence the environment at various political or administrative levels; also to show how sensitive different parts of the landscape are to human driving forces at local, regional and national level. The reason for the complexity of landscape systems needs further investigation to verify that more simple analytical approaches will not suffice. A likely outcome is that landscape systems are complex due to the fact that social, economic, technical and ecological processes operate on a wide range of spatial and temporal scales.

Remote sensing and GIS modelling have been used effectively for study and mapping of agrolandscapes. Indeed the availability of up-to-date spatial information about land use and land cover is an indispensable condition for successful modelling. Aerial sampling, aerial photography, aerial video and satellite images are tools for producing conventional, objective land-survey maps. Because indigenous systems of land classification are based on observable characteristics, remote sensing can play a useful role. The use of remote sensing, combined with a ground-sampling strategy, yields an accurate and credible product.

Traditional thematic maps (land use, soils, vegetation, etc.), integrated into a GIS of agrolandscapes, have been used in this study as a source of information for a database. Nevertheless, these maps have a number of shortcomings that may cause spatial inaccuracy in any future model. These include:

(1) The data on these maps may be outdated. Though all national level maps of Russia are updated at least once in 10 years, the reliability of the data is arguable. The large-scale survey of landscapes and land use, especially of the regions of Siberia that are difficult to access, requires tremendous political and financial support that could not be given during the period of economic transformation. Also, land-use and even landscape structure have considerably changed in recent years, following the transformation of agriculture and the economy.

(2) Many of the maps which will be integrated in the database were created for educational purposes (Map of Agricultural Regionalization, Map of Land Use). Their main feature is that the classification of landscapes, and accordingly regionalization, is carried out 'from above': natural zones are determined according to their solar energy income, the sector division is based on humidity conditions, and provincial and local peculiarities are used for more detailed mapping. This approach is extremely efficient for understanding the nature of land-cover stratification, but information may be lost in marginal areas due to the specific generalising approach.

(3) The scale of human impact on the present-day landscape structure is not yet clearly determined. It is possible that anthropogenic disturbance has, in some cases, changed the zonal structure of natural landscapes. If so, efforts must be made to create an updated picture of this zonal structure, because almost all maps of natural resources and land use in Russia are based on the existing borders of natural zones.

(4) As far as vegetation cover is concerned, no reliable cartographic information exists for almost the entire area of the former Soviet Union. The maps available at the moment mostly show potential stratification of the vegetation cover and not the present-day situation.

The problems mentioned above could be partly resolved by use of remotely-sensed data. Coarse, medium and high resolution imagery from several satellite systems (NOAA AVHRR, RESURS-01-3/MSU-SK, RESURS-F/MK-4) have been processed, interpreted and analysed. The high frequency of AVHRR data coverage makes it possible to monitor change in land cover conditions (Miller *et al.*, 1988; Justice *et al.*, 1985; Goward *et al.*, 1985). Particular attention has been given to the analysis of vegetation indices (NDVI – Normalised Difference Vegetation Index), which are very efficient for mapping terrestrial vegetation and land-cover patterns.

RESULTS OF SCALE-DEPENDENT AGROLANDSCAPE RESEARCH

Country scale

Experiments with the vegetation index show that the traditional scheme of vegetation and land-cover zones, developed by Dokuchaev (1899) and implemented in different studies for a variety of agrolandscape zonation schemes, needs to be re-examined. Here, two main dimensions of the existing scheme are analysed: the structure of the natural zones and their transformation under human impacts, including agriculture.

Structure of natural zones

According to the seasonal dynamics of vegetation and biomass, the zones identified in this study differ from the existing stratification of forest landscapes in the north of Eurasia. Only two zones based on the humidity of the landscapes should be identified: (a) an extra-humid forest zone that mainly corresponds with the existing zone of the northern taiga, but also occupies a part of the tundra and central taiga; and (b) the zone of forests with optimal humidity that corresponds with the zone of the southern taiga and mixed forests. Both of the newly determined zones show a fixed sub-latitude appearance throughout almost all Russia (which cannot be said about the traditional zones). Possibly, in the south of the forest zone, step zones should also be stratified, based on the amount of humidity and warmth (normal humidity, dry, extra dry). These zones do not correspond strongly to the existing structure and have a better sub-latitude appearance than existing zones: they are based on the dynamic of energy and water input, which is expressed in the vegetation dynamic and not in the floristic characteristics of land cover.

Anthropogenic transformation of the structure of natural zones (including agricultural transformation)

Experiments show that agriculture and other types of human activity have, in some cases, changed the appearance of natural zones. One of the areas of active anthropogenic influence that may be seen and analysed by remote sensing techniques is the area north of the Black Sea near the Dnieper and Kuban rivers, an area of intensive artificial irrigation. The vegetation activity in this humid steppe area differs a great deal from the dry steppe of the Volga region and West Siberia, though traditional vegetation and land-cover maps combine them in one zone. The use of remote sensing data allows us to determine a new scheme for natural zones, taking into account anthropogenic impact.

Regional scale

Data from two different sources of satellite images were used for the regional scale study – vegetation index data with 10 km and 1 km resolution and high resolution false-colour composite photographic images. Cluster classification methodology was used for data analysis. Data from each source give different information about the pattern of vegetation and land use: vegetation index data are most efficient for the study of the seasonal phenological dynamic of vegetation and space photographic images are most important for analysing long-term trends of vegetation and land use.

Vegetation index data have been used to delineate natural and economic landscape units at a regional scale as well as land-use types. Each land-use type can be distinguished by its unique seasonal vegetation change. Territories with a high percentage of arable land, for example, have lower vegetation index values in March than forests, because larger areas are covered with snow. In the centre of the European part of Russia, regions of intensive agriculture may be distinguished on this basis (e.g. the arable land area near the city of Vladimir). The images of July and August were used to stratify arable land according to the type of crops (cereals do not grow in these months, which leads to low values for the vegetation index; but potatoes and vegetables still grow). Each landscape unit or land-use type has its unique vegetation index pattern that can be used for description and analysis. On the basis of the vegetation index images for different months of a year, we may determine each landscape unit or land-use type; if this information is available for different years, it gives us the opportunity to study long-term dynamics of vegetation and land use in the region. This principle was used for land-use/land-cover stratification and analysis of seasonal dynamics.

False-colour photographic images allow a new approach to be taken towards land-use/land-cover classification and definition of natural and economic districts. Experiments in case study regions of European Russia have shown that land-use maps produced by automatic classification of remote images differ a great deal from existing vegetation and landscape maps. New maps are much more detailed and exact in describing land use and landscape stratification. Different mathematical methods were used to check the results of automatic classification of remotely sensed data coefficients, describing spectral and spatial complexity of resulting classes; the basic methods of automatic image recognition were implemented for automatic determination of economic and natural districts.

Traditional ground research was undertaken to augment results from remote sensing and to resolve problematic items. At the regional scale for Central Russia, 5 major cluster categories were determined using 1 km seasonal vegetation activity data from AVHRR by applying the unsupervised classification described above. By visual interpretation and the analysis of attribute data and case studies, these clusters were primarily defined as the following land-cover types:

(1) Water bodies,
(2) Agricultural land,
(3) Deciduous forests,
(4) Coniferous forests,
(5) Non-vegetated areas.

The image containing these clusters was then analysed using traditional cartographic and text data about land-cover particularities of the study area as well as on land use. In general, this analysis revealed the landscape structure of the study area to be composed of several natural–anthropogenic regions, with specific combinations of cluster pattern for each region.

By a comparative analysis of land-cover data from traditional and remotely sensed sources for each of these natural–anthropogenic regions, the objectives below were achieved.

Specification of geographical and landscape meaning of defined cluster categories

The cluster 'Agricultural Land' comprises arable land under cereals, pasture and meadows. However, widespread 'Non-vegetated areas' clusters in the south of the study area, for example on the Middle Russia Upland, indicate that some of the garden and potato production areas fall into this cluster. Moreover, some of the agricultural land falls into the 'Coniferous Forests' category, especially on the watersheds of the southern part of the area where there is highly intensive agriculture. Here, pine forests are rare but appear on the sandy terraces along the larger rivers. This cluster also appears in the poorly forested areas of the Dneper–Desna and Meshera lowlands and contributes to the landscape meaning of the 'Coniferous Forests' cluster (see below).

The 'Deciduous Forests' cluster combines the smallleaf forests that are widespread in the north with the broadleaf forests preserved in the south. In addition, the cluster includes almost all of the mixed spruce/smallleaf forests shown on currently existing vegetation maps as pure spruce forests. This cluster is dispersed on the Valday Upland and west of Moscow – these areas are described on existing maps as continuous spruce forests. Furthermore, in the Bryansk region this cluster includes oak–pine–spruce forests on sandy river terraces. Thus the geographical interpretation of the 'Deciduous Forests' cluster category is relatively dry, well-drained environments with young forested vegetation cover.

The 'Coniferous Forests' cluster identifies pine forests very clearly over most of the territory, except for the Bryansk region. However, the analysis of ground truth data had shown that the patterns of this cluster on the image do not correspond directly with forests on traditional maps; they include extensive wetlands covered with thin pine and birch–aspen forests that are classified on traditional maps as non-forested territories. This cluster also includes, as described earlier, some agricultural land in almost non-forested southern areas. Apparently, this cluster comprises relatively wet, poorly-drained environments that are covered with pine forests, wet environments with thin forests in the north and centre of the study area, and non-forested, wet agricultural land.

The 'Non-vegetated areas' cluster is comprised of urban territories: all the major cities are clearly defined on the image, for example Moscow, Tula, Tver, Yaroslavl, Vladimir, Bryansk, Kaluga and Volgograd. In addition, some agricultural land falls into this cluster, mainly garden and vegetable production areas. This is demonstrated by the widespread distribution of this cluster on the Oka–Don Upland and around large urban agglomerations – the so-called 'garden belt'. It should be noted that determination of these territories as 'non-vegetated' may be caused by the specific dates of data collection: in April crops are not yet growing; in June crops are not as developed as the natural vegetation and cereals; by September the crops are already harvested.

Thus, we may refine a description of the selected cluster categories as follows:

(1) Water bodies,
(2) Dry agricultural land,
(3) Dry forests,
(4) Wet forests and agricultural land,
(5) Poorly vegetated urban and rural land.

Correction of published cartographic data on land cover from results of this study

The comparative analysis of classified images with ground truth data had shown that, over most of the study area, patterns of major land-cover types coincide. This allows us to conclude that the classification is reliable for regional studies of land-cover structure and dynamics. However, there are sites where cluster patterns differ from available data from traditional sources.

Cluster images have several significant qualities: they are compiled by an impersonal computer technique that is the same for the whole study area; and they show up-to-date information. Compared with this, traditional maps are highly influenced by subjective factors related to the procedures of ground data collection and generalisation. Moreover, the information on traditional maps can be decades old. This means that cluster composite images may be used successfully to verify and correct the landscape meaning of land-cover categories on traditional maps.

Comparing traditional forest maps with composite cluster images, some significant shortcomings can be identified:

(1) These maps often show thin forests as non-forested areas due to the low economic value of the timber.

(2) Contours on traditional maps often do not show a real correspondence to forests and non-forests due to manual generalisation algorithms; they more often correspond with variations in soils, relief and land use and thus show the landscape characteristics of the territory.

(3) Traditional forest maps often distort species composition to improve economic value of forests. Due to this, many mixed forests are marked as purely coniferous, while thin forests on wetlands are not even counted as forests.

(4) Even the most recently verified hand-written maps are outdated by decades, especially in relation to the economic restructuring of agriculture.

The time gap between real changes in agriculture and land cover and the output from traditional mapping processes explains the problems with topographic and land-use maps, and thus of landscape and ecological maps as well. For example, these maps do not trace the growth of brushwood on abandoned arable land in areas where farmland is degrading. Also, traditional maps do not measure degradation of forests cover related to industrial exploitation and agricultural expansion. Nor is the continuing incursion of urban, transport and industrial land uses onto arable and forested areas clearly identified.

It is evident that these limitations of traditional small-scale maps can be overcome by analysis of a time series of remotely-sensed data classified using the proposed technique.

Determination of major tendencies of land-use/cover dynamics in agricultural regions

The combined analysis of traditional and remotely sensed data allows us to determine the following, sometimes contradictory, patterns of land-use/cover dynamics:

(1) Urban development on agricultural land;
(2) Spontaneous recovery of forests;
(3) Anthropogenic recovery of forests;
(4) Controlled deforestation around urban agglomerations and transport routes;
(5) Changes in species composition of forests (through clearance of valuable timber);
(6) Waterlogging of arable and forest lands next to wetlands.

All the processes as mentioned above, together with industrial and agricultural pollution and overlaid by the diversity of natural and anthropogenic processes, form the dynamics of land use/cover; they may be investigated, to varying degrees, by classification of a time series of space imagery.

Local scale

Two case studies in Central Russia (Moscow and Tver regions) have been analysed in detail using false-colour images of very high resolution. Preliminary experiments suggest that these two case studies illustrate contrasts in the way that nature and economic activity correlate in landscape structure.

In the case study of the Moscow region, the differentiation of natural landscapes is reinforced by anthropogenic (mostly agricultural) activities, and land-use structure is highly correlated with natural landscape structure. Ground study data will be used for further analysis of this thesis.

In the Tver region study area, natural landscape differentiation has been eliminated by long-term intensive agricultural activity. Here the collective form of agriculture in the Soviet period, especially economic specialisation of different parts of the territory, was not sympathetic to the structure of natural landscape. Some of the natural characteristics of the region, such as the widespread distribution of loam soils and the homogeneous relief, promoted the process of masking the structure of the natural landscape. For these reasons, vegetation and land cover developed according to the specialisation of local farms. Each of the farms created its own natural–agricultural region, largely ignoring natural landscape structure.

AGROLANDSCAPES OF THE CENTRAL EUROPEAN PART OF RUSSIA, TRANSITION PROBLEMS AND THE PROSPECTS OF ECOLOGICAL AGRICULTURE

Russia has a very rich natural agricultural capacity and a large internal market, but recent changes in the Russian economy have put severe strains on agricultural production. In particular, the collapse of the kolkhoz system has resulted in widespread degradation processes effecting land cover. While the ideas of an ecologically sound agriculture were developed by famous Russian geographers many years ago, they could not be implemented in the former social system.

Russia is nowadays serving as a huge research laboratory where changes in landscape structure under the influence of human driving forces, especially within agriculture, can be studied. The combination of socialist and capitalist forms of land use, and the differing development of the regions, has resulted in a great diversity of favourable and unfavourable land-use management practices. The analysis of the landscape character of the European part of Russia provides tremendous experience for development of environmental policies and practices.

A local case study is being carried out to investigate the potentiality for such an analysis. The study area is situated in Central Russia (the Moscow Region, 100 km west of Moscow). The main agricultural specialisation is breeding of dairy cattle within a highly developed farming system of potato and forage crop cultivation, but set in a peri-urban location. The present high impact of agriculture on peri-urban landscapes is mostly due to the high proportion of row crops, the large amount of pesticides in use and soil damage. The important factors for the agrolandscape and ecological processes are contained in natural physical conditions (physical environment) and land-use practices (socioeconomic environment).

One of the most important production units within the region is the 'Borodino' kolkhoz. It has a major land-use conflict with the Military Historic Museum Reserve. Important events of two wars took place here, so that the area of protected landscapes is 567 sq. km and the area of regulated building is 645 sq. km. This is much more than the total area of the kolkhoz. Up to now, all problems have been resolved by mutual agreements between the two administrations. In general, the kolkhoz management board is ready to concede in reasonable cases (e.g. when the Museum can prove that a cattle breeding farm is polluting a river and should be removed). But lately, the administration of the Museum has been demanding that all the land of the kolkhoz belonging to the Reserve should be abandoned. Although it is agreed that agriculture could still develop in the region, the Museum's administration now acts as if it were the only land owner, with all local farms in the area subordinate to it.

At the end of 1992, 80 people left the kolkhoz after obtaining their own land (approximately 30 ha each). Most subsequently united into farmer associations according to their type of specialisation rather than location of farmland. About 15 farmers did not join any association and have their own agricultural businesses. In future these farms are unlikely to be very productive because of financial and technical weakness. Nevertheless, the 'Borodino' kolkhoz still exists as a legal entity: it owns a fodder processing plant, a dryer and a food plant in Mozhaisk city.

Following the landscape methodology developed previously, a map of agrolandscapes has been created for the study area. The next step in landscape evaluation involved matching requirements of the land-use types (LUT) with the properties of natural land(scape) – NLU. In this way, the suitability of the NLUs for relevant LUTs has been determined. Mostly, the NLUs are suitable for more than one LUT. Therefore, the results of the land(scape) evaluation in combination with a number of land-use options have been presented to the local administration to facilitate decision making.

The local level of land-use management has become more important under the developing market economy. The most important tasks which must be resolved at the local level include:

(1) Land use planning and forward projection;
(2) Setting aside areas for different kinds of land use, including farms, recreation zones, personal ownership and building;
(3) Definition of differential taxes for land use;
(4) Definition of differential rates on land for private ownership;

(5) Establishment of the most favourable land-use conditions regarding ecological circumstances and economic efficiency.

To solve these and many other problems of land-use management, it will be important to assemble all kinds of information concerning quantitative and qualitative characteristics of land use in the region. This information will be used to support the decision-making process for the whole region as well as for smaller areas. For example: in the border areas of settlements, it may help to plan future building and to determine the land suitability for different types of land use from the natural and socioeconomic points of view; or when making a decision about leasing areas covered with forests, the decision maker must have information about the legal status of the forest, the quality of wood, suitability for different kinds of land use: afforestation or felling, recreation or personal ownership.

The local landscape/land-use data collected throughout 10 years of interdisciplinary research has been used to create a database for the case study area. The software chosen for the practical implementation of the programme is IDRISI (4.0). The computer system is being used for modelling and spatial analysis. An essential land-use assessment task has been to identify and map all the sites suitable for a specific economic activity. In this way, the extensive databases collected during field-based research, together with interpretation of remotely sensed data for the landscapes of the European part of Russia, allow us to carry out analyses of the spatial distribution of present day landscapes, in particular agrolandscapes, and to build simple models to be used in environmental decision making and prediction of possible trends. Since Russian agriculture, as described earlier, is in the process of economic transformation, the most important issue for predictive modelling will be the analysis of land use and landscape structure changes caused by economic processes, in other words, how the economy may influence the natural subsystem of a landscape.

CONCLUSION

This chapter has shown how the landscape concept could enrich assessment of the land use/cover situation and help in delineating landscapes as a complex system with different patterns of relationships. Together with quantitative computer methods, the landscape concept may serve as a scientific base for long-term development of ecologically sound and economically efficient agriculture in Russia. Such agriculture will help to use natural potential more effectively and avoid the negative consequences of land degradation.

Many geographical distributions, such as soil variables, are inherently complex, revealing more information at higher spatial resolution apparently without limit. Since a computer database is a finite, discrete store, it is necessary to sample, abstract, generalise or otherwise form complex information. 'Geographical data modelling' converts complex geographical reality into a finite number of database records or 'objects'. The objects are represented by points, lines or polygons, and also possess descriptive attributes.

However, a universally acceptable database on land use is not an easy undertaking, either in conceptual or computer programming terms. Such a database can only function if it is part of a wider system of related databases of land resources and crop requirements and is used in a framework with clearly defined societal goals and constraints. We have emphasised the need for a quantitative analysis of the biophysical aspects of land suitability as a starting point. The next stage would be an assessment of the socioeconomic feasibility and acceptability of options for land-use development and their impact.

The landscape concept in Russia must be placed in the context of the present transition to a market economy and democratic governmental practices. Under these processes, the principles and methods of nature management are being transformed, while implementation of land reform and farm restructuring is progressing. The agricultural economy is becoming more profitable due to more efficient land-use management and the reduction of post-harvest losses. In this context, landscape planning can become an important method of agricultural management, helping to determine the set of natural and socioeconomic constraints in which any farmer or decision maker acts.

Changes in agriculture play a most significant role in transformation of the spatial distribution of landscapes and land use. Though changes in the industrial and urban infrastructure may have tremendous influence on the economy and lifestyle of the countryside, from the spatial point of view they do not usually cause remarkable changes at regional and country levels. Such changes usually take place inside existing spatial objects (polygons, lines or points) or occupy insignificant areas next to these objects. From this point of view, analysis and modelling of spatial patterns of agriculture is of vital importance for understanding the existing situation and possible trends. The agrolandscape approach, when implemented by the GIS technique, has proved to be very efficient.

REFERENCES

Dokuchaev, V. V. (1899). Teaching about nature zones. Sankt-Petersburg, 28 pp. (In Russian). (Kucheniyu o zonakh prirody.S.P., 28 pp)

FAO (1976). *A framework of land evaluation.* Soil Bulletin 32, Wageningen

Forman, R. T. T. and Godron, M. (1986). *Landscape Ecology.* Wiley, New York, pp. 620

Goward, S. N., Tucker, C. J. and Dye, D. G. (1985). North American Vegetation Patterns Observed with the NOAA-7 Advanced Very High Resolution Radiometer. *Vegetation,* **64**: 3–14

Justice, C. O., Townshend, J. R. G., Holben, B. N. and Tucker, C. J. (1985). Analysis of the Phenology of Global Vegetation using Meteorological Satellite Data. *International Journal of Remote Sensing,* **6**: 1271–318

Milanova, E. V., Kushlin, A. V., Middleton, N. and Soyuzkarta, M. (eds) (1993). *World Map of Present-Day Landscapes.* Soyuzkarta, Moscow

Miller, W. A., Howard, S. M. and Moore, D. G. (1988). Use of AVHRR Data in an Information System for Fire Management in the Western United States. In Proceedings, *International Symposium on Remote Sensing of Environment,* 20th, Nairobi, Kenya, December 1986, Vol. 1, pp. 67–79

Skole, D. (1992). Scientific requirements for a 1 km data set. In Townshend, J. R. G. (ed.) *Improved Global Data for Land Applications.* IGBP Report No. 20, Stockholm, pp. 11–23

Stomph, T. G. and Fresco, L. O. (1991). *Describing agricultural land use.* A draft. FAO, ITC, Wageningen Agricultural University

Tappan, G. and Moore, D. G. (1989). Seasonal Vegetation Monitoring with AVHRR Data for Grasshopper and Locust Control in West Africa. In Proceedings, *International Symposium on Remote Sensing of Environment,* 22nd, Abidjan, Cote D'Ivoire, October 1988, Vol. 1, pp. 221–34

Turner II, B. L., Moss, R. H. and Skole, D. L. (eds). (1993). *Relating Land Use and Global Land-Cover Change.* IGBP Report No. 24, Stockholm, pp. 65

CHAPTER 12

ECOLOGICAL–ECONOMIC PROBLEMS OF LAND USE IN THE UKRAINE

V. Voloshyn, S. Lisovsky, I. Gukalova and V. Reshetnik

INTRODUCTION

The Ukraine is the second largest European country by area (60.4 million ha) and sixth in population size (52.2 million). Its territory is divided into forest, forest steppe and steppe zones. Based on their ecology and economy, 9 regions can be identified (see Figures 12.4 and 12.5), each with characteristic land-use structures and specific environmental and economic problems.

The country's main natural resource potential lies in its land resources. Occupying more than 0.46 percent of the Earth's continental area, the Ukraine has the largest share of the world's black soils: 2.2 percent on a global scale and 23.2 percent on a European scale (World Resources, 1990:268–69 and National Economy, 1992:339). Minerals are the second most important natural resource, among which the reserves of coal, iron and manganese ores stand out.

These two natural resources – soils and minerals – have underpinned the development of agriculture and heavy industry, respectively, and both sectors are significant in the economic structure of the country. These two dominant natural resources have also shaped the character of land use in the Ukraine but have been modified in different ways by global and historical forces.

THE CURRENT STRUCTURE OF LAND USE

The major part of Ukraine's territory (95 percent) is lowland. The only mountains are the Carpathians and the Crimea, situated in the far west and south, respectively. Nearly three-quarters of the territory is occupied by the forest steppe and steppe zones.

Today, almost all of the country's territory is in economically productive use. For example, 70 percent of the Ukraine is used for agriculture, 55 percent of which is arable (crop) land: these figures are higher than in most other European countries. Only Hungary has similar indices (Statistics, 1992:10).

Nearly 8.2 percent of the territory is used for non-agriculturally productive purposes. Almost 1.14 million ha are occupied by water reservoirs and other hydrological infrastructure, the majority of which are situated in administrative regions near the Dnipro river. The 0.69 million ha of the Dnipro cascade water reservoirs alone represent 60 percent of total water reservoir capacity. Almost 1 million ha are occupied by infrastructure for transportation, and 0.18 million ha are damaged and waste land. The latter are found mainly in Dnipropetrovsk (33.3 thousand ha) and the Donetsk region, where the main enterprises of the mining industry are concentrated.

Nearly 3.3 percent of the land area is in urban use. The Donetsk and Lugansk regions are the most urbanized and densely populated among the administrative regions of the Ukraine, urban land making up 6 percent and 5.7 percent, respectively, of their total areas.

In the Ukraine, the share of technogenically transformed land, including built-up areas, open-pits, open-cut mines and other mining areas, is greater than in most other European countries (Figure 12.1). Also, the density of population is above average for Europe, although transport infrastructure and second homes cover a relatively small area compared with other countries. Consequently, relatively few landscapes can be considered close to 'natural'. For example, forest covers less than 15 percent of the Ukraine, which is half the average of both the world and European situations. Indeed the area of forest decreased by almost a half during the nineteenth and twentieth centuries. On water, the area occupied by artificial reservoirs is greater than that of natural lakes and rivers. Nevertheless, only a small proportion of the area of the Ukraine is under legislative protection: five times less than in the world and 10 times less than in Europe (Figures 12.2 and 12.3) (World Resources, 1990:300).

ANTHROPOGENIC TRANSFORMATION OF TERRITORY AND DEGRADATION OF ENVIRONMENT

Excessive economic utilization of Ukrainian territory has caused almost complete destruction of natural ecosystems, especially in the forest steppe and steppe zones. Thus, a significant depletion in the range of species has occurred and a considerable number of plant and animal species are close to extinction. The total degradation of more than 22,500 small rivers is one of the consequences of the territory's anthropogenic transformation. Hydrological conditions at micro and meso levels have changed almost everywhere and this, in turn, has aggravated the entire hydrological system.

The predominance of arable farming in the land-use structure is a particularly negative factor within the general syndrome of excessive anthropogenic transformation. Arable farming destroys both self-regulating mechanisms in the environment and the stability of natural ecosystems, giving no opportunity for the development of a balance between arable, grass and forest areas.

One of the main factors accelerating the process of environmental destruction has been intense urbanization of the last 20–30 years. Cities, with their concentration of economic production and population, are now the most ecologically affected locations within the Ukraine. They have become generators of billions of tons of solid waste and particularly harmful industrial sewage. Urban transport is also a factor having a significant negative impact on the environment. Evidence of the ineffectiveness of controls over the

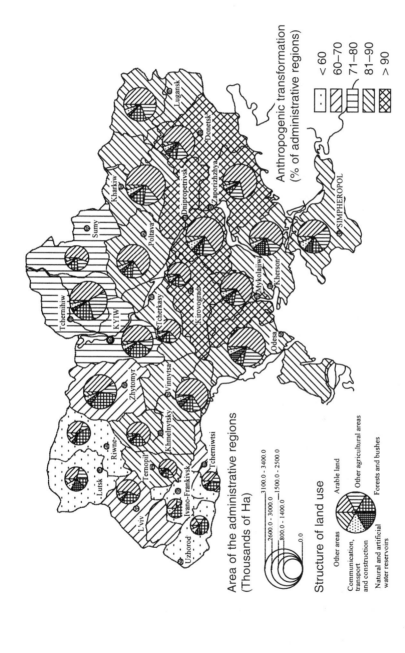

Figure 12.1 Anthropogenic transformation of the Ukraine

1

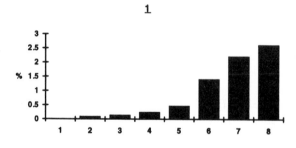

2

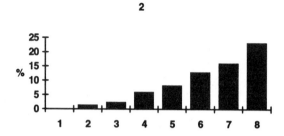

1. Territories which preserve their natural status
2. Reserves and territories under protection
3. Water resources reserve
4. Territories covered with forest
5. Total area of the territory
6. Full amount of water use
7. Area of arable lands
8. Grain production

Figure 12.2 The Ukrainian share of some global (1) and European (2) ecological and production indices

economic system as regards protecting natural ecosystems can also be found in areas for leisure and recreation.

Finally, one of the biggest technogenic catastrophes in the world, the nuclear reactor accident at Chernobyl, has had an impact on the entire character and development of land use in the Ukraine. As a result of this accident, an area of more than 5 million ha (almost one tenth of the Ukrainian territory) has a level of radioactive nuclide pollution of more than 1 Ci per sq. km.

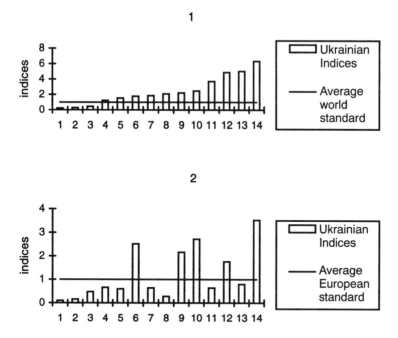

1. Ratio of land under protection to total area
2. Water supply
3. Ratio of forested areas to total area
4. Yield of cereal crops
5. Use of fertilizers per hectare of arable land
6. Emissions of sulfuric oxides (calculated per pure sulfur) per US dollar of gross domestic product
7. Water consumption per square kilometre
8. Production outputs per square kilometre
9. Availability of arable land per inhabitant
10. Total electricity consumption in the economy as ratio of electricity power production output to size of gross domestic product
11. Emissions of sulfur oxides per square kilometre
12. Ratio of arable land
13. Electricity production output per square kilometre
14. Intensity of use of water reserves (ratio of amount used to amount of resources)

Figure 12.3   Some indices to characterize intensity and efficiency of land use in the Ukraine as compared with the world (1) and Europe (2) (standardized)

THE IMPACT OF AGRICULTURE ON STATE AND QUALITY OF THE LAND RESOURCE

As mentioned previously, agriculture, as the major land use in the Ukraine, has played an important part in the anthropogenic transformation of the country as a whole, resulting in degradation of the land resource. Inappropriate cultivation practices have damaged soil structure and promoted soil erosion. For example, soil is damaged by widespread use of heavy agricultural machinery.

At present, 43 percent of the arable land in the Ukraine is threatened by soil erosion and, for the most productive black soils, this figure rises to 70 percent (Figure 12.4). In the large administrative districts of Vinnytsya, Tcherkasy, Kirovohrad, Kherson and Mykolajiw, more than 75 percent of the surface area is affected. For the Ukraine, between 1961 and 1990, the area of lightly eroded soils increased by 26 percent, while the area of medium and highly eroded soils increased by 23 percent. Every year, 24 million tons of humus, nearly 1 million tons of nitrogen and 0.7 million tons of phosphorus are washed-out from the fields (Status, 1991:15).

The high acidity level of soil is a particularly serious problem caused by use of fertilizers, sludges of industrial origin and excessive irrigation. The proportion of such land in the Ukraine reaches one third of total arable area. The majority is situated in the Vinnytsya, Tcherkasy, Kirovohrad, Tchernihiw, Sumy, Trans-Carpathian and Ivano-Frankivska regions (Ukraine, 1993:20).

Negative environmental processes are also observed in areas subjected to hydrological modification, which in the Ukraine cover approximately 6 million ha, including 3.2 million ha of drained and 2.6 million ha of irrigated land. Nearly 14 percent of the irrigated arable land has suffered erosion, 5 percent is over-watered, 7.7 percent is soured, and more than 30 percent is saline and saliferous. The proportion of such soils is especially high in the Kherson region and on the Crimea (Status, 1991:16–17).

The major part of the drained land is concentrated in the Ukrainian Polissya territory. Here, during recent years, soil erosion and deflation risks have increased by 27 times. Forty-three percent of the drained black soils have increased acid levels, 18.4 percent are subject to wind erosion, and 4.6 percent are exposed to water erosion (Status, 1991:17).

The use of agrichemicals and organic and mineral fertilizers is one of the major factors responsible for pollution of soils and surface and subsurface water. Nevertheless, the amount of mineral fertilizers per hectare is only 60 percent of the average in Europe (Figure 12.3), or 2.1 times less than in France (World Resources, 1990:280–1). The total quantity of pesticides used in 1993 was 71 thousand tons or nearly 2.2 kg per hectare of crop land. The largest amount of pesticides is used in the Vinnytsya, Tcherkasy, Tchernihiw, Volyn regions and in the Crimea (Information Bulletin, 1994:141).

ATMOSPHERIC POLLUTION AND ITS IMPACT ON LAND USE

One of the major anthropogenic factors having a negative impact on land use in the Ukraine is air pollution (Figure 12.5). In 1993 the total amount of pollutant emissions into the atmosphere was nearly 10 million tons. Emissions from stationary sources reached 7.3 million tons and those from transport 2.7

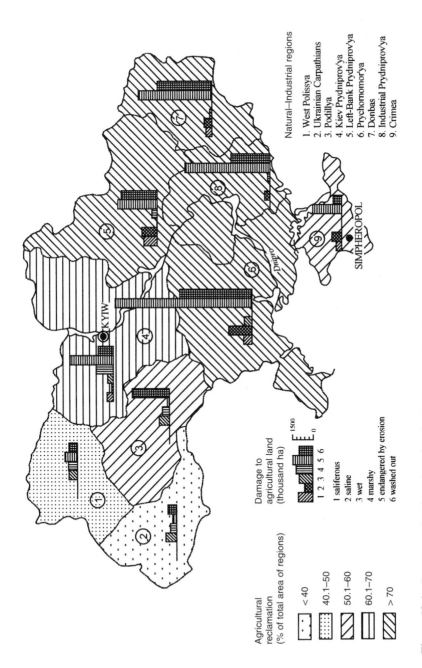

Figure 12.4 Damage and reclamation of agricultural areas of the Ukraine

Natural–Industrial regions

1. West Polissya
2. Ukrainian Carpathians
3. Podillya
4. Kiev Prydniprov'ya
5. Left-Bank Prydniprov'ya
6. Prychornomor'ya
7. Donbas
8. Industrial Prydniprov'ya
9. Crimea

Damage to
agricultural land
(thousand ha)

1500
0

1 saliferous
2 saline
3 wet
4 marshy
5 endangered by erosion
6 washed out

Agricultural
reclamation
(% of total area of regions)

< 40
40.1–50
50.1–60
60.1–70
> 70

KYIW

SIMPHEROPOL

Dnipro

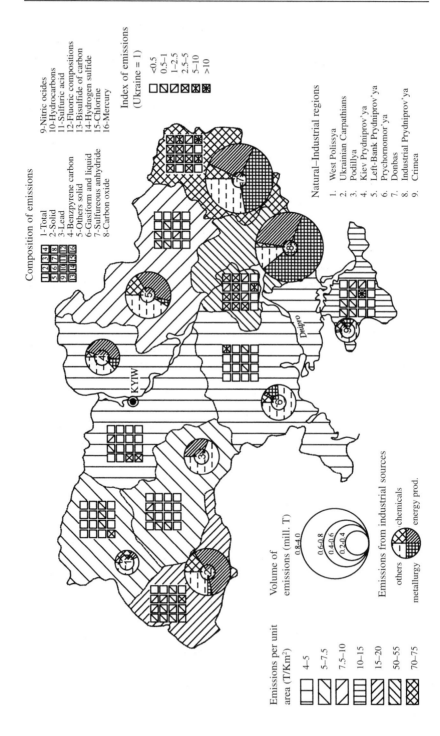

Figure 12.5 Industrial emissions of harmful substances into the atmosphere in the Ukraine

million tons. In comparison to 1992, the total amount of pollutants decreased by 2.2 million tons (pollution from stationary sources decreased by 1.3 million tons and from transport by 0.9 million tons). This decrease is due to the economic crisis and the sharp recession in industrial production.

The essential components of emissions from stationary sources are sulfurous anhydride and carbon monoxide. In 1992, carbon monoxide accounted for 30 percent (2.6 million tons) of the total amount of pollutants emitted into the atmosphere. Sulfurous anhydride emissions amounted to 27.5 percent (2.38 million tons) of all emissions, while emissions of ash, dust and other solid particles accounted for 20 percent and nitric oxides for 7.4 percent (0.64 million tons) of the total.

The main components of transport emissions are nitric and sulfuric oxides, with volatile organic compounds and lead being the most harmful.

In 1992, the Ukrainian share of global atmospheric pollution was 1.7 percent (*Acid News*, 1994:14). Calculated per sq. km, the average emission volume was 3.7 times the world's average and contributed 70 percent to the total volume of European emissions. For nitrogen emissions, the situation is similar.

Heavy industrial enterprises are the largest sources of atmospheric pollution. Concentrated in the Donbas, the industrial Prydniprov'ya and on the Crimea, these enterprises emit more than 30 percent of the total amount of wastes: manganese (97.7 percent), mercury (91.3 percent), chrome (81.7 percent), lead (75.2 percent), arsenic (34.2 percent), hydrochloric acid (29.9 percent), nickel (25.5 percent) and sulfuric acid (12 percent).

Second place is occupied by factories of the Ministry of Energy (thermal power stations). Their share in the total amount of pollutants comes close to 30 percent: nearly 63.8 percent of all sulfurous anhydride emissions and 57.3 percent of the vanadium emissions. Among the 76 European industrial enterprises representing the largest sources of sulfur emissions into the atmosphere, there are 12 Ukrainian thermal power stations. The power stations of Krivoriz'ska and Burshtinska are placed 14th and 15th respectively (*Acid News*, 1994:3).

Seventeen percent of the total emissions from stationary sources come from the coal mining industry and 4.6 percent from the chemical and petrochemical industry. The latter produces 24.2 percent of the total amount of emissions of volatile composition, 53 percent of ammonia, 80 percent of sulfuric acid, 57.7 percent of arsenic, 36.2 percent of nitric acid and 34.4 percent of hydrochloric acid.

Atmospheric pollution is especially heavy in the largest cities, namely in the industrial centres of the Donetsko–Prydniprovsky region. Among them, the regions of Kriviy Rig, Dnipropetrivsk, Donetsk, Dniprodzerzhinsk, Kostyantinyvka, Mariupol and Zaporizhzhya have the highest rates of pollution. As a result, the territories adjacent to these cities are experiencing considerable environmental damage. These areas are mainly used for intensive agricultural production and their agricultural products – mainly vegetables, milk and dairy products – have substantial contents of harmful chemical residues. In addition, the Ukrainian forests experience considerable degradation resulting from atmospheric pollution.

THE PROBLEMS OF WATER RESOURCES

There are severe problems in relation to the provision and use of water resources. The Ukraine has a particularly low capacity as regards its water quality, reaching 28 percent of the world standard and 17 percent of the average European standard (UN, 1994:37). On the other hand water consumption is very high: calculated per sq. km, the consumption of water is 1.85 times higher than average world water consumption. The intensity of water use in the Ukraine (i.e. the ratio of water consumption to reserves) exceeds the world level by 6.3 times and that of the European average by 3.6 times (Figure 12.3) (World Resources, 1990:167). In this respect, the Ukraine holds third place in Europe after Belgium and Bulgaria, while the index exceeds that in the USA by 2.9 times and France by 4.75 times (Statistics, 1992:15). In consequence there is a shortage of water, with surface and subsurface water resources experiencing depletion and pollution.

The country's water resources mainly originate in the Dnipro river. Nearly 70 percent of the Ukrainian population uses Dnipro water, while 15 percent use the water of other rivers. The needs of the remaining 15 percent of the population are met by subsurface water. The main consumer of water is industry, with 53.1 percent of total water use. Thirty-one percent of industrial water is consumed by electricity power plants, 8.3 percent by the metallurgy sector and only 13.5 percent by other branches of industry. Thirty-six percent of water is used by agriculture, while 11.1 percent is used by the population and municipal economy.

There is an inverse relationship between the amount of water used and the water reserves of particular regions in the Ukraine. Most of the water is used in the south where it is scarce; the resources consist of water transported from the Dnipro, Yuzhny Boug and Dnister Rivers and small reserves of subsurface water. Water consumption reaches its highest level in the Dnipropetrivsk, Zaporizhzhya, Donetsk regions and in the Crimea. At the same time, the northern and southern parts of the Ukraine are well supplied with water resources. They belong to regions with a relatively low level of water use. Thus, in 1993, more than 25 percent of the total amount of water used was consumed in Dnepropetrivsk and Zaporizhzhya regions, while another 16 percent was used in the Donetsk and Lugansk regions.

The water resources of the Ukraine are subject to intense pollution from a number of different sources, although industry is responsible for large-scale pollution. In mining areas, subsurface depression water cones have come into existence as a consequence of intensive drainage of highly mineralized and open-pit waters and their storage in surface water reservoirs. In certain regions (i.e. Krivorizhzhya, Donetsk and Lugansk regions), these cones cover vast areas.

The chemical and metallurgy industries appear to be the worst contaminators of water. The enterprises of the chemical industry are widely spread throughout all regions of the Ukraine and their impact is clearly visible. They affect the quality of water of the Dnipro River, and its associated reservoirs, and the coastal area of the Azov Sea in the area near Maruipol. The waters most heavily polluted by chemical enterprises are the Siversky Donets river in Lisytchansk city area, the Carpathian rivers, as well as the territory and coastal area of the Black Sea and the Northern Crimea.

Another industrial activity having a strong negative impact on water resources is the electricity generating industry. After the construction of the Dniprovsky cascade water reservoirs, as well as a series of hydroelectricity power stations (HPS) and their associated water reservoirs on other rivers, reservoirs have become accumulators of a variety of harmful emissions from economic activities, for example radioactive nuclides, heavy metals and washed-out pesticides, herbicides and mineral fertilizers. In 1990 the quantity of caesium 137 in the Kiev water reservoir amounted to 7200 Ci, while in the Cremenchug water reservoir it was 294 Ci (Information Bulletin, 1994:102). The water reservoirs in the Dnipropetrivsk and Zaporizhzhya regions are to a large extent polluted by emissions from ferro-metallurgical enterprises.

As a result of an increasing evaporation rate in the area of HPS water reservoirs, there is an increasing lack of water. In the summer period, almost all the water reservoirs of the Ukraine which dry out experience a process of efflorescence, while adjacent areas are flooded and become boggy.

Thermal and nuclear power stations cause severe thermal and radioactive pollution of the water in their cooling systems. Another negative industrial impact on water resources is the pollution of surface and subsurface water resulting from the infiltration of liquid wastes from sedimentation tanks and slime pits. Water pollution also occurs as a result of transport activities, including loading and unloading activities in ports and accidents in oil pipelines.

Agriculture contributes to water pollution mainly through washed-out organic and mineral fertilizers and agrochemicals, as well as through manure from intensive animal production plants.

Lastly, there is intensive water pollution from cities as a result of poorly developed infrastructure, technical breakdowns and obsolete technology in the municipal economy. This has a particularly negative impact on the coastal areas of the Black and the Azov Seas, thereby endangering their ability to recover from environmental damage.

This catalogue of facts suggests that optimizing water use and protecting hydrological resources are key problems, the solution of which is vital if positive changes in land use in the Ukraine are to be achieved.

THE STRUCTURE AND STATE OF THE ECONOMY AS THE MAIN CONSTITUENT OF CRISES IN LAND USE

One of the main reasons for the current critical state of land use in the Ukraine is that for a long time the country has been developing highly intensive agricultural production as well as heavy industry, particularly mining, power generation, metallurgy and chemicals. However, the anthropogenic impacts on the environment and the intensive exploitation of natural resources have not been compensated by a high value of final gross domestic product (GDP): calculated per sq. km of territory, the GDP amounts to less than one third of the European average.

Ukrainian agriculture is characterized by low productivity, calculated as output either per hectare of land or per unit of labour. The average yield of grain in recent years, for instance, is only 70 percent of the average in Europe (Figure 12.3), 45 percent of that in The Netherlands, and 54 percent of France (World Resources, 1990:278–9). The productivity of labour in Ukrainian agriculture is 4.6 times lower than in France, while production per hectare is 2.8 times lower (Voloshin *et al.*, 1994:105). It should be noted that such a standard of agricultural production is achieved on black soils which, until recently, were the world's reference for fertility.

Ukrainian industry is amongst the most resource-consuming and ecologically insecure in the world. At the beginning of the 1990s, 1.2 times more energy resources were needed than in Russia to produce one dollar of GDP, 1.6 times more than in Belorussia, and over 10 times more than in west European countries. At the same time, the production of one dollar of GDP caused an output of nitrogen oxides twice as high as the world average and 3 to 4 times higher than in other European countries.

The current state of the land reserve and environment in the Ukraine provides a convincing demonstration that 'neglecting environmental protection issues over many decades has caused a degradation of considerable areas of the country, with so-called transitive economies as well as their inability to ensure economic activity in long-term prospects' (UN, 1994:14).

At present, there are no analogous territories in the world of an area comparable with the Ukraine: none has the same high concentration of nature-consuming and resource-spending branches of the economy which, at the same time, have such a low standard of economic efficiency. Today, under the conditions of economic crisis, the Ukraine's structurally distorted economic complex does not supply practical economic opportunities to ensure a real improvement of environmental conditions. This, in turn, is further aggravating negative economic conditions.

CONCLUSIONS

To ensure conditions for sustainable development of the Ukraine, a deep restructuring of the national economy and its technological complexes is necessary, along with a decrease in consumption of natural resources by the economy and an optimization in use of natural resource potential. The government, acting within scientifically-based programmes, could intervene to achieve recovery of the natural environment of the Ukraine. Solutions are needed which aim to halt the process of environmental degradation and restore the self-regulating capacity of the environment.

Allowing for further social and economic developments, there is a need to achieve an effective relationship between agriculture and land transformed into ecological reserves, including recreational areas, protected from strong anthropogenic influences. There is a need to ensure an optimal relationship between damaged lands, natural meadows and pastures, and forested areas. Small rivers in the Ukraine need to be recreated and the hydrological structure improved. A complex of nature protection interventions is particularly needed in the following regions: Donetzko-Prydniprovsky, Ukrainian Prichornomorye (lands adjacent to the Black Sea), Polissya and the Carpathians.

These strategic interventions will only be successful if they are based on new scientific approaches to the organization of land use. The resulting land-use structure would be optimized at local, regional and state levels, taking into account their natural resource characteristics. Such a long-term strategy would have to be placed in the context of both the further evolution of market relations in the Ukraine and global processes in the division of labour.

At the same time, attention is needed to the optimization of agricultural production. Here, the objective is to decrease anthropogenic stress on the environment to a level which does not exceed critical ecological limits. Measures would have to be introduced gradually to restore sound environmental conditions on which human activity could be based. As the first step in this strategy, a decrease in the total area of arable land by 20 percent of the present area should be sought, so as to prevent damages to the banks of small rivers and the erosion of slopes in fields. New woodland strips are needed to protect fields from erosion, coupled with changes to optimize both the shape and area of fields. Additional important objectives are extending the forested area up to 25 percent of the total land area, and increasing the area of land under protection to between 6 and 8 percent of the total.

Addressing these issues would help to ensure stable social and economic development in the Ukraine and, at the same time, resolve the urgent problem of increased environmental protection.

REFERENCES

Acid News (1994). N 5, December 1994, Goeteborg, Sweden

Industrial Statistics Yearbook 1991 (1993). United Nations, New York, pp. 948

Information Bulletin (1994). *Information bulletin on the status in the geopolitical environment of the Ukraine during 1992 to 1993*. Release 13. Kiyv, pp. 160

National Economy (1994). *National economy of the Ukraine in 1993. Statistics yearbook*. Kiyv, Tehnika, pp. 494

Nello, S. S. (without year). The food situation in the ex-Soviet Republics. *Soviet Studies,* **44**(5): 857–80

Statistics (1992). *The environment in Europe and North America.* Annotated statistics 1992, United Nations, New York, pp. 336

Status (1991). *Status of land resources and ecological consequences of productive power development.* Ukraine's SA, Kiev, SOPS, pp. 22

Ukraine (1993). *Natural environment and the humans.* Series of maps, Kiyv, pp. 55

UN (1994). *General Assembly. Development and international economic co-operation.* General Secretary Report, pp. 56

Voloshin, V., Gukalova, I., Lisovsky, S. and Mischenko, N. (1994). Ecologo-economic features of agricultural land use in the Donetsko-Prydniprovsky region of Ukraine. In: *Analysis of Landscape Dynamics – Driving Factors Related to Different Scales.* Leipzig, pp. 98–110

World Resources (1990). *World Resources 1990–91.* New York, Oxford: Oxford University Press, pp. 383

CHAPTER 13

QUALITATIVE CHANGES IN NATURAL LANDSCAPES
AND LAND USE FOLLOWING RADIOACTIVE
POLLUTION FROM THE CHERNOBYL ACCIDENT

V. Davydchuk

INTRODUCTION

The Chernobyl nuclear accident, which took place at the nuclear power plant
(NPP) on April 1986, occurred in the northern part of the Ukraine, in the
Polesie woodland. Following an explosion in the power unit, a great amount
of radioactive material was thrown out of the damaged nuclear reactor and
spread into the environment. The surroundings of the NPP were heavily
contaminated with the radioactive isotopes. This made normal human life and
activities impossible and caused significant changes in the natural landscapes
and land-use system of the territory.

PHYSICAL CONDITIONS AND THE STRUCTURE OF NATURAL
LANDSCAPES

Polesie is a fluvioglacial sandy plain on the northeastern slope of the
Ukrainian Crystalline Shield in the area of the Dnieper stage of the quaternary
glaciation, which corresponds to the Saale stage in Western Europe.

Climatic conditions

The climate of Polesie is temperate semi-continental. It is characterized by
moderately cold winters and moderately warm summers: the average
temperature of the coldest month (January) is about −5 to −6°C. The average
temperature of February is higher, although February is a month of extreme
temperature minima (−22 to −25°C, sometimes −33 to −35°C). In March and
April there is a rapid rise in temperature (6–8° per month). Late April and
early May are periods of the last frosts: frosts on the surface of the soil can
occur up to late May. July is characterised by both maximum average annual
temperatures (17–19°C) and high temperature extremes (35–36°C). In autumn
the decrease in temperature begins in the second half of September and
continues during October and November – more than 6°C per month. The first
autumn frosts typically occur at the end of September or at the beginning of
October. Thus, the period without frosts lasts for 150–170 days.

The season of vegetation growth begins in the middle of April and continues up to the end of October: its duration is about 220 days. The sum of the
day-degrees above 10°C is about 2660°. The average annual precipitation is

about 580 mm, with variations between 400–850 mm. Up to 70 percent of the annual precipitation occurs in the warm period of the year.

Relief and landscapes

The Polesian region is formed mainly by a plain with low relief (110–145 m above sea level). A number of different natural landscapes can be identified based on patterns of relief, surface deposits, soils and vegetation; among these are river flood plains and terraces, terminal moraine ridges, moraine–fluvioglacial and limno–glacial plains. This part of the Ukraine is known for its considerable swampiness. In the accident zone, bogs account for 22.5 percent of the territory.

The landscape of the Chistogalovka terminal moraine ridge, which stretches from the city of Chernobyl to Chistogalovka village, represents the upper level of the relief of the Chernobyl NPP 30-km zone. The hills of the ridge reach 260 m above sea level. They are composed of fine, fluvioglacial sand, bedded with a push moraine of sandy loam at a depth of 0.4–1.0 m and palaeogenic clays and sands at a depth of 15–30 m. The soils are sod–podzolic dusty–sandy (pH 4.5–4.9, humus 1–2 percent). In the pre-agricultural period the ridge was covered by forests with a predominance of *Quercus robur* L., *Fraxinus excelsior* L. and *Carpinus betulus* L., which are now substituted by planted wood stands of *Pinus silvestris* L. origin, secondary forests of *Betula pendula* Roth. and former agricultural land which is now covered by long-term fallow grass vegetation.

Another terminal moraine ridge of small size is situated between the villages of Kamenka and Opachichi, at the southern border of the evacuation zone. Its northwestern part is a low, ripple moraine–fluvioglacial plain which lies 15–35 m lower than the terminal moraine ridge. This level of the relief is composed of fluvioglacial sand, bedded by a morainic loam at a depth of 0.8–1.0 m. The share of particles smaller than 0.05 mm in diameter is very low (4–8 percent). This explains the poorness of the sod–podzolic sandy soil formed here (pH 4.2–4.5; humus 0.8–1.5 percent). The vegetation is mainly pine forests.

A number of isolated depressions with diameters from 0.3 to 0.8 km are scattered on the surface of the landscape. Most of them are occupied by swamps with alder forests or sedge–reed bog coenoses.

The landscapes of the Dnipro and Prypjat terraces are very typical of the Chernobyl zone. The flat surfaces of the terraces are composed of alluvial sands with a deficiency of clay and small particles (4–6 percent). Thus, the sod–podzolic sandy soil formed here is dry and poor (0.4–0.6 percent of humus). The terraces are mainly covered by planted pine forests of different ages. The rear, lower parts of the river terraces are occupied by eutrophic bogs with peat bog soils (the thickness of the peat is about 0.5–2.5 m) and forests of *Alnus glutinosa* (L.) Gaertn., or by sedge–feed bog coenoses.

The flat surface of the terrace between the Dnipro and the Prypiat is drained by the Braginka river and its small tributaries, together with a dense network of drainage channels and ditches. These small streams have very large (1–2 km) flood plains which are composed of peat of 0.5–1.5 m thickness, with peat bog soils. Before the Chernobyl accident, most of these organic soils had been drained and cultivated. Now they are covered by a dense long-term fallow grass vegetation. Non-drained, 'natural' peat bog soils are to be found with the alder forests and sedge–reed associations.

In the streams, channels and bogs, there are a large number of dry sandy islands. The soils of these areas belong to the sod–podozolic type with a very thin (8–12 cm) humus horizon. Particularly poor sod–podzolic sandy soils, which have about 0.5–1.0 percent of humus in the upper horizon, are covered by different types of natural pine forests. The more fertile sod–podzolic sandy soils were formerly occupied by agricultural land.

The most typical forest ecosystems of the region are *Pinetum cladinosum* and *Pinetum phodococco-dicranosum,* with a ground cover of *Koeleria glauca* (Sorend.) DC, *Festuca sulcata* (Hack.) Num.p.p., *Antennaria dioica* (L.) Gaertn., *Phodococcum viti-idaea* (L.) Avror., *Dicranum scoparium* Hedv., *Cladonia silvatica* (L.) Hoffm. and *C. rangiferina* (L.) Webb.

The lowest level of the relief of the Chernobyl accident zone is represented by the landscapes of the flood plains of the Dnipro and Prypjat rivers and their tributaries. The flat surfaces of the front and middle parts of the flood plains are dissected by a number of meander lakes and channels, and composed of layered alluvial sands and loams. They are covered by shrubs of *Salix acutifolia* Willd., with a grass–xerophytous ground cover and sedge–hygrophytous meadows on alluvial soils.

Ecosystems and types of land cover/land use

The main ecosystems and types of land cover/land use in the Chernobyl accident zone are shown in Figure 13.1, together with the background of soil contamination by Cs-137. The figure shows that the forests of the region mainly consist of different types of *Pinetum*. The total area of *Pinetum* in the accident zone covers more than 30.5 percent of the territory of the 'wet triangle', or about two thirds of its forested area. Because of regular forest farming during the last 130 years, the woodland stands of the territory mainly belong to young and medium age groups (Figure 13.2).

Polesie is known for the considerable swampiness of its river flood plains and terraces. In the accident zone, bogs account for 22.5 percent of the territory. Two thirds of the bogs were drained during the decades immediately before the accident and were employed as farmland, accounting for 40 percent of the arable area. The location of the agricultural land is shown on Figure 13.1.

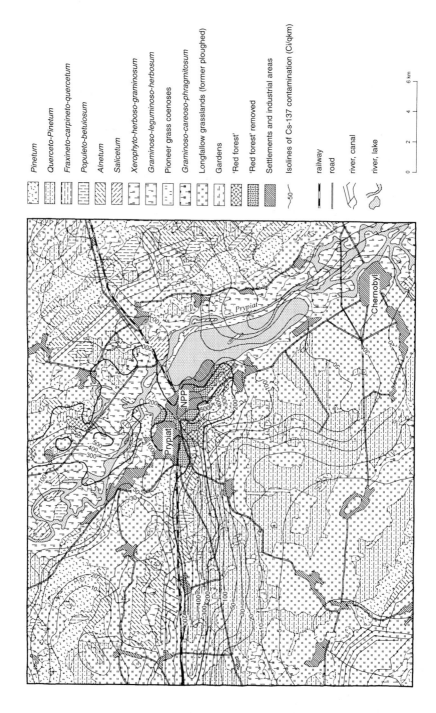

Pinetum

Querceto-Pinetum

Fraxineto-carpineto-quercetum

Populeto-betulosum

Alnetum

Salicetum

Xerophyto-herboso-graminosum

Graminoso-leguminoso-herbosum

Pioneer grass coenoses

Graminoso-careoso-phragmitosum

Longfallow grasslands (former ploughed)

Gardens

'Red forest'

'Red forest' removed

Settlements and industrial areas

Isolines of Cs-137 contamination (Ci/qkm)

railway

road

river, canal

river, lake

Figure 13.1 Land use of Chernobyl accident zone

238

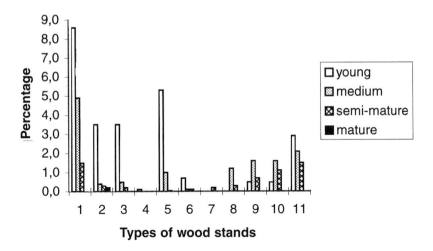

Figure 13.2 Types of wood stands and the age structure of the forest ecosystems of the Chernobyl accident zone (% of total area).
1. *Pinetum xerophyto-cladinosum;* 2. *Pinetum rhodococco-dicranosum.* 3. *Querceto-pinetum coriloso-graminoso-herbosum;* 4. *Querceto-carpineto-pinetum euonimo-coriloso-aegopodiosum;* 5. *Pinetum vaccinoso-polytrichosum;* 6. *Carpineto-quercetum euonimo-coriloso-aegopodiosum;* 7. *Carpineto-quercetum coriloso-galeobolonoso-aegopodiosum;* 8. *Fraxineto-carpineto-quercetum coriloso-geumoso-filipendulosum;* 9. *Pineto-betulosum herbosum;* 10. *Populeto-betulosum vaccinoso-polytrichosum;* 11. *Alnetum higrophytosum*

Thus, the variety and heterogeneity of the landscapes and ecosystems of the Chernobyl accident zone are related to the diversity and structure of land-cover/land-use types.

LAND-USE CHANGES AND SPONTANEOUS LANDSCAPE RESTORATION AFTER THE ACCIDENT

Historically, the structure of land use in the Polesian region of the Ukraine has been related to natural conditions. Agriculture was characterised by intensive farming (rye, potatoes, flax, vegetables), while livestock breeding made use of the natural forage resources of the river meadows. Before the accident in 1986, arable land accounted for 38 percent of the territory of the present evacuation zone; 10.5 percent was occupied by meadows, 36 percent by forests and 4.5 percent by settlements (Figure 13.3).

Land-use changes

The Chernobyl nuclear accident caused irreversible radioactive contamination of densely populated areas. Normal human life and land use became

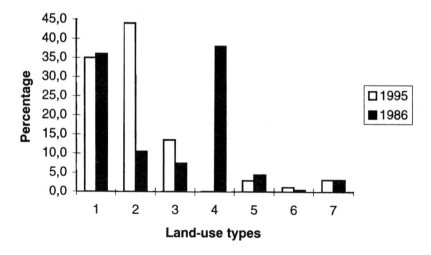

Figure 13.3 Land use in the Chernobyl accident zone in 1986 and 1995 1. Forest, 2. Meadows and grassland, 3. Swamp, 4. Arable land, 5. Settlement, 6. Industrial area, 7. Water

impossible on the territory contaminated with more than 40 Ci km^{-2}; 90,784 persons, including 61,614 people from the two towns of Prypjat and Chernobyl, and 29,170 of the rural population of 78 villages were evacuated. However, more than 3775 km^2 of Ukrainian territory, with a population of 253,100, were contaminated by more than 5 Ci per sq. km of Cs-137 but are still inhabited and cultivated (Table 13.1).

Some agricultural counter-measures, such as liming, incorporation of absorbents into the soil, special ploughing and fertilising, were carried out to improve the radiological situation after the accident. Nevertheless, 2115 km^2 of agricultural land, including 1352 km^2 of arable land, were lost for reasons of high contamination or high transfer coefficients to agricultural products from the acid peat soils. Consequently, a further 48,500 people were evacuated from 80 settlements in addition to those who had been evacuated immediately after the accident.

The land-use system of the evacuation zone had to be adapted to the new conditions. About 580 ha of pine forest in the immediate neighbourhood of the nuclear power plant were lost by lethal irradiation (500 mR/h and more) during the first months after the accident. Because of the colour of the dry pine needles it was designated as 'red forest'. Most of it was removed mechanically, together with the surface layer of contaminated soil, in order to decrease the dose rates for NPP staff. As a consequence, the 'industrial area' as a type of land use increased after the accident (Figure 13.3).

Some abandoned villages have been burned since the accident by spontaneous fires or deliberately destroyed so as to reduce the probability of fires and the secondary resuspension of the radionuclides. Thus, the percentage of

Table 13.1 Contamination, area and population of the inhabited territories of the Ukraine

	Contamination Ci/km²		
	5–15	*15–40*	*>40*
Territory	2355 km²	740 km2	680 km²
Population	204,200	29,700	19,200

'settlement area' in the evacuation zone decreased from 4.5 percent to 3 percent of the territory between 1986 and 1995.

Normal agricultural activities within the evacuation zone were stopped completely and only some experimental arable fields remain. This has resulted in the quite rapid and spontaneous (self-restoration) revegetation of land that had been cultivated previously.

Spontaneous restoration of landscapes and ecosystems

The spontaneous restoration of landscapes and ecosystems in the evacuation zone includes the succession of natural vegetation, recovery of the wild animal population, renewal of the initial structure of the soil horizons, including the chemical soil parameters of acidity and humus content, and removal of pesticides and artificial nitrogen from former arable land.

These features are developing partly in the context of natural processes and partly human activity (measures to counteract radioactivity and zone management). As a result of the evolution of the ecosystem, the conditions of migration of radionuclides are changing. The model of plant succession in the ecosystem is based on a forest typology (Figure 13.4); this typology in turn is based on data from investigations of landscapes and ecosystems of the region, as well as from studies in the Chernobyl zone before and after the accident. Long-term fallow grass associations, for example, replace the agrocoenoses after agricultural activity has ceased. The long-term fallow grass shows a predominance of *Elytrigia repens* (L.) Nevski and *Oenotera biennis* L., with occurrence of *Achillea millefolium* L., *Crepis tectorus* L., *Hypericum perforatum* L., *Festuca ovina* L. (in dry habitats), *F. rubra* L., and *Potentilla argentea* L.

The first stage of succession – the segetal herb community – is followed by a grass-dominated community and a shrub stage of succession. The present ruderal grass community represent the second phase of succession on the former arable land of the Chernobyl 30-km zone; there is a predominance of *Setaria glauca* (L.) Beauv., *Chenopodium album* L. and *Erigeron canadensis* L., which dominated the abandoned arable land of the evacuation zone during the first years after the accident. Thus, spontaneous succession of natural vegetation consists of renewal of the forest coenoses typical of the Ukrainian Polesie.

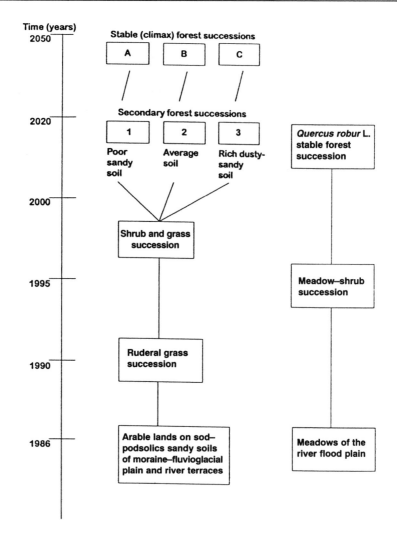

Figure 13.4 Plant succession in the evacuation zone of the Chernobyl nuclear accident

CONCLUSION

Since the nuclear accident of April 1986, ten seasons have passed and the natural landscapes and ecosystems continue their evolution. Arable land has been replaced by long-term fallow grasslands so that the grass-covered areas have increased considerably (Figure 13.3). Secondary forest successions on the former arable land and stable (climax) forest successions in the existing forest ecosystems are expected to develop within the next 25–75 years. Of course, human intervention and forest fires can accelerate the processes or stop them. Using the landscape map and models of succession, forecasts of the evolution of ecosystems and landscapes can be developed as the basis for evaluation of migration of radionuclides, change in the radioecological situation and application of technologies of decontamination.

The natural evolution of abandoned landscapes creates new, inconstant structures of land use on this territory. This will minimise the surface migration of radionuclides out of the evacuation zone. Thus, the importance of ecological factors for the improvement of the radiological situation need to be taken into consideration in elaborating a strategy for the management and rehabilitation of the abandoned zone.

CHAPTER 14

CONCLUSIONS

R. Krönert

In this publication, the working groups and members of the MAB project present and extend the results and insights obtained over the years that they have been working on the project. In their contribution, Reenberg and Baudry reiterate the underlying purpose. An important aspect has been the comparison of landscape pattern dynamics under various natural, as well as cultural, conditions. This comparison is necessary if we are to understand the causes and ecological consequences of landscape changes across the diverse range of European agricultural regions. It has become clear that there can be no one fixed methodological approach for such comparative studies, at least not one that fits everyone's needs. There are several reasons for this. The main reason is that natural and socioeconomic conditions in rural areas differ considerably throughout Europe. There are also a wide range of agricultural enterprise structures. Additionally, project members come from various disciplines, each bringing their own approaches and working methods. This diversity is reflected in the contributions to the book, and it is this multidisciplinarity which proved of such value in the mutual learning process which accompanied project participation and which is bound to be of long-term value to project participants. It is also important to point out that there were differences in the analytical data available for each land and region and this certainly has an influence on research approaches and project results. The net outcome is that each contribution to this book is capable of standing alone, and yet each still manages to express the sense of common purpose running through the project. A universal goal was the search for methods which can bring agricultural production into closer harmony with its natural environment, and which can identify those socioeconomic conditions required for such change. The research contributions are also united in the challenge to minimise the environmental damage caused by agriculture and to improve diversity of the countryside in the interests of broad-based nature conservation.

A few basic principles concerning a methodological framework must, however, be highlighted. As Reenberg and Baudry point out in this publication. 'It must be concluded that such a framework must encompass a variety of scales – from the field level (where techniques are applied to perform a certain land use), to the household level (where decisions are taken), to the regional and country level (where policies are implemented). The focus on scale and on spatial aspects of phenomena and processes is central. Time-series analysis is a natural part of studies which deal with landscape dynamics. In this context too, the selection of scales will deserve considerable attention'.

Rural areas can and should be seen and studied as agricultural landscapes, as economic and living areas, and as planning areas, each made up of different hierarchical levels. In doing so, physical, biological and socioeconomic components and processes, as well as developments and development trends over a long period of time need to be taken into account. Key research objectives and the definition of the spatial unit will differ according to the context in which a rural area is considered. In all cases, it is the real, existing landscape that is observed, what Milanova *et al.* (Chapter 11) call the present day landscape. If we want to understand the dynamics of these landscapes, then we need to look at the driving forces behind landscape changes. These are generally related to socioeconomics or planning. However, actual changes in rural land use and the rural landscape take place through actions of the farmer (or other land user). For methodological reasons, the study of agricultural landscapes can, however, focus on the study of natural complexes and identification of natural potential, as recommended by Bastian in Chapter 3. Land use can be more closely examined – research may go as far as to map land use on individual fields over a period of several years. Such land-use orientated research can also include mapping of small biotopes, or study of land-use change over a period of several decades. Another key topic for more intensive research is the flow of materials and energy in the landscape. Where the rural area is studied as a living and economic area, then central attention may be given to agricultural enterprises and their economic behaviour, to land use down to the level of individual fields, to family and age structures, education levels, the purchasing and sales environments for agricultural inputs and products, price levels (for agricultural products, machinery and fertiliser), rural settlement patterns, agricultural support measures and to agricultural policy. Of course, attention must also be given to the natural conditions for agricultural production, and to the interactions between the natural complex, agricultural production and rural social structures. If rural areas are treated as planning areas (see Bacharel and Pinto-Correia, Chapter 4), then management units often form the basis of research. Land-use plans are prepared and models used as a formulation for development objectives. These models contain, among other things, positional statements on land use, agricultural structural development, and nature conservation. Plans are based on analyses of natural conditions, current land-use structures, enterprise structures, etc. From the approaches sketched out here, it is clear that the rural area is treated as a system which can be broken down into system components. Each component is then investigated in detail, also in terms of its interactions with other components. Dependent on the research objectives and the system functions tackled, there is clearly a variety of possible methods for studying rural areas; this is also reflected in the contributions within this book.

The work of the MAB project group was dedicated to a subject area which is of fundamental importance both to European integration and to regional development. Focus was placed on land use and land-use changes, and on their natural and socioeconomic causes. The ecological consequences of land use also received considerable attention. It is interesting to note that these issues are also the core subjects of meetings of the International Association of Landscape Ecology (IALE). Just as remarkable is the fact that a common IGBP/IHDP Core Project on land use and land-cover change (LUCC) exists. This project covers both the natural and social sciences and is based on the recognition that, 'the contemporary state of the world's land cover is a constantly changing mosaic of cover types determined by both the physical environment and human activities. These changes in land cover can have profound global consequences' (Skole, 1996). Despite these and other research activities involving changes and developments in European rural areas, there is still an urgent need for a systematic study of the development of agricultural regions as individual agricultural landscapes, as economic and living areas and as planning areas. A typology of European agricultural regions would provide a valuable resource on which future studies of the dynamics and development potential of different types of agricultural regions could be based.

REFERENCE

Skole, D. (1996). Land-Use and Land-Cover Change: an Analysis. *Global Change Newsletter,* **25**: 4–7

INDEX

abandonment   33
  Chernobyl   239–241, 243
  farm land   95–96, 98–99, 115–118, 239
    Normandy   105–106, 112
Achillea millefolium L.   241
acid rain   24, 25, 182, 208
Acts
  Act on Private Farming, Estonia (1989)   170
  Environmental Basis Act (1987) 65–66
  Land Reform Act, Estonia   170
  Main Act on protected areas   65, 67
  Nature Conservation Act (1937)   94
afforestation
  see also marginalization
  as an indicator   95
  erosion amelioration   13, 14
  pattern   32–33
  positive outcomes   131
  programmes   192
  scheme, Poland   196
agglomerations   141–163
Agricultural Statistical Yearbook (1993) 195
Agricultural and Forestry Map, Portugal 73
agricultural machinery   31, 92, 226
agriculture   3
  see also farms and farming
  growth   206–207
  impact on land resource quality 225–226

intensification   195
land use   121, 143–145
landscape   2–4, 12–17, 159–160
state intervention   8, 121–139
agroforestry   35, 76–77
agrolandscapes   206, 207–218
Alentejo   65–79
Algarve   66
Alnus glutinosa (L.) Gaertn.   236
analysis and diagnosis   43–63
  see also frameworks; modeling; spatial analysis
  cluster analysis   107–108, 212–215
  comparative   23–25
  data   165–166, 245
  hierarchical   26–27, 118
    ecological systems   32, 52, 104–105
    landscape evaluation   9
    methods   50
    order   46, 50
  indicators   54–56, 58, 60–63, 133–136
    afforestation as   95
    animals as   55
    assignment of   50–51
    biotopes as   90
    climate as   51
    land use seen as   12
    landscape objects as   48
    principle   46–48, 63
    species as   10, 56
    variables   44
  landscape changes, time aspects 50–53

material and methods 14, 53, 165–166
prediction 14
scales 24, 27, 36, 38, 48–50, 55
time series analysis 29, 37, 37–38, 215, 245–246
universal algorithm 48
animals
  see also habitat; livestock; species
  communities 200
  corridors for 90
  as indicators 55
  legal protection 66
  migration 59
  populations 142, 179, 241
  species 15, 55, 60, 177, 179
Antennaria dioica (L.) Gaertn. 237
Arable Area Payments Scheme (AAPs) 124, 129
arable land 9, 13, 15, 84, 98, 141, 143–144, 170, 221
Arable Payment Areas 124
Arhus 85
AVHRR 212
Avror 237
Azov Sea 231

Bad Lauchstadt 1
Baltic-German landlords, Estonia 168, 173
barriers 11–12, 183, 202
Belgium 230
Belorussia 232
Betula pendula Roth. 236
biodiversity
  biological 9
  biotic regulation potential 55
  complexity in rural and agricultural landscapes 14
  improvement 11
  influence of landscape structure on 200
  loss 6, 23, 53
  management 17
  planning for and management 16–17
  preservation 8–9, 185
biofuel production 131
biogeochemical barriers 11–12, 202

biomass production 47
biosphere cyclic attributes 208
biotic regeneration potential 46, 47, 61
biotic regulation 46, 51, 55–60, 63, 148
biotopes
  age 59
  diversity 94
  dynamics 91–94, 99–101
  endangered 60
  grassland 62
  hierarchical order of 50
  as indicators 90
  linking 56, 58
  mapping 17, 55, 56, 246
  networks 9
  protection 66, 148, 161
  secondary 151
  spatial variations 93
  values 60, 62
birds
  see also species
  eagles 142
  great bustard 72, 75–78
  Important Bird Area (IBA) 75–76
  population increase 177–179
  species 10
  stork 142
  wild geese 151
Bitterfeld/Wolfen chemical industry 150–151
Black Sea 211, 231
bogs 59, 90, 92, 93, 167, 177
Braginka river 237
Brandis, Leipzig 146
Brittany 32
Bryansk 213–214
Bulgaria 230

Cadastre Book of Estonia and Livonia (1918) 165
Caen 1
Calvados 34, 106, 117
CAP *see* Common Agricultural Policy
capitalism 4, 104–105
carabids 107
Carpathian Mts 2, 13, 221, 226, 233
Carpathian river 231
Carpinus betulas L. 236

cartographic data, correction of
    published data 214–215
Castro Verde 71, 72–78
cattle 87, 106–118, 169, 171
    see also livestock
census data 106, 125
cereals 11, 106, 122, 160, 225
Chaplowski, General Dezydery 193
chemical industries 13, 150–151
Chenopodium album L. 241
Chernobyl 224, 235–243
Chistogalovka village 236
Cladonia silvatica (L.) 237
Clays 131–136
climate
    as an indicator 51
    change, global 25
    conditions, Chernobyl 235–236
    as a driving force 28, 96
    Leipzig-Halle 154–155
    Mediterranean 70
    meteorological data 166
    Pays d'Auge 106–107
    Wielkopolska 189–190, 198–199
coenoses 47, 236, 241
*Collaborative Programme on Application
    of Remote Sensing in the Management
    of Less Favoured Areas* 17
collectives *see* cooperatives
COMECON countries 166
Commission of the European
    Communities 65, 123
Common Agricultural Policy (CAP)
    see also set-aside
    application 77–78
    ecological consequences 8
    nature conservation 65
    reforms 84, 99, 121, 123–125, 236
        Regulation 2078/92 125
    structural policies impact 72
compensation areas 183–184
conservation areas 15, 55, 66, 135, 185
contamination
    see also pollution
    groundwater 7, 142, 146–148, 160,
        185
    material flows 9, 11–12, 14, 246
    radioactive 224, 235–243
    soils 159, 237

water 6, 159
cooperatives 1, 6, 7, 93–94, 169,
    170–171
    liquidation 144, 154, 216
Copenhagen 85–88, 90–94
CORINE Land Cover Map 17
corridors 10, 35–36, 60, 66, 90, 183
Coturnix coturnix 10
Countryside Access Scheme 130
Countryside Stewardship Scheme 130
Cremenchug water reservoir 231
Crepes tectorus L. 241
Crimea 226, 229, 230, 231
Crimea Mts 221
cropland
    crop production 31, 84, 91, 98, 122
    cropping methods 91
    crops 73–74, 87, 105–107, 143
    forage 107, 109
    irrigated crops 73–74
    loss of 32
    rotation 6, 13, 74, 84, 98, 143, 150,
        191–192, 195
    Russia 206–207
Czechoslovakia 1

Dahlener 142
database systems 27, 33, 35, 219
    IDRISI software 218
decontamination 46, 243
degradation
    arable land 13
    environment 12, 52, 222–226, 232,
        233
    forests 230
    land cover 216
    landscape 23
    soils 14, 194, 201
Denmark 1, 3, 12, 17, 32, 38, 81–102,
    125
Dessau 141
Diocranum scoparium Hedv. 237
diversification 103–119, 121–123, 133,
    137
Dnieper river 21
Dnieper-Desna lowlands 213
Dnipro river 221, 230–231, 236–237
Dniprodzerzhinsk 229
Dnipropetrovsk 221–222, 229, 230

Donbas 229
Donetsk 221–222, 229, 230–231
Donetsko-Prydniprovsky region 229, 233
drainage
  decrease in biotope value 60
  land use changes through 193–194
  landscape changes through 30–32, 46, 177
  leaching of phosphate ions through 11
  marshlands 6, 193–194
  pipes/ditches 90, 92–94
  practices, Wielkopolska 199
  subsidies 94
  valleys 189
Dresden 46, 61
driving forces 28
  agriculture 4–5
  climate 28, 96
  Common Agricultural Policy (CAP) as 103
  increase in yields 195
  land use 28–29, 81–119, 142–145, 167–172
  landscape 24, 54
  policies 82–83
  socioeconomy 99
  technology 93, 99–101
Dübener Heide 142
dynamic activity monitoring 17
dynamic processes 105
dynamics
  affecting biotopes 91–94, 99–101
  of diversity 176–177
  landscapes 81–102
  of the share of agricultural land, Estonia (1918–1992) 175
  of species pattern 10

Earth Summit Rio (1993) 138
East Anglia 127
ecological development 205–220
ecology
  analysis and diagnosis 36–37, 43–48, 52
  consequences of Leipzig-Halle-Bitterfeld agglomeration 145–148
  evaluation 48, 50

indices, Ukraine 224
landscape as an ecological system 105–106
landscape state 50, 53–55, 60–61, 63
Leipzig-Halle 155–160
networks 167, 182–184
patterns 103, 104
planning projects 55
research and planning 44–45, 48
systems 16
economy 9, 23, 99, 105, 107
  changes 54
  conditions 54, 98
as a driving force 28
efficiency 15
ecosystems 8, 26–27, 35, 55, 59–70, 201, 237–239
Elbe river 142
Elytrigia repens (L.) 241
Emberiza calandra 10
emissions 225, 226, 228–230, 229
England 121–139
Entradas 72–73
Environmental Basis Act (1987) 65–66
environmental goods 129–130
Environmental Protection regional authority of Saxony-Anhalt 155
Environmentally Sensitive Areas (ESA) 84, 129–130, 136
Erfurt-Halle railways 154
Erigeron canadensis L. 241
Estonia 1, 8, 165–189
Estonian Agricultural University Institute of Environmental Protection 166
Estonian Enterprises Register 170
Estonian Land Use Cadastre (1939) 165
Estonian and Livonian Bureau for Land Culture 173
Estonian Meteorology and Hydrology Institute (EMHI) 166
eucalyptus 71, 74, 76
EULANU 15
EUROMAB 1–21
European Commission green paper, Perspectives for the CAP 123
European Community (EC) 75, 106

European Ecological Network
(EECONET) 165, 185
European Standard (UN, 1994:37) 230
European Union (EU)
Nitrate Directive 91/676/EEC 130
Regulation 3887/92 124
agricultural policy 4–6, 25, 65, 84,
94, 99
directives 160
fallow regulations 7, 144
grants 148
post-productivist transition 121–139
reform measures (1992, 1998) 124
Regulation 1094/88 127
reports 77–78
Environmental Protection regional
authority of the State of Saxony 155
extensification 15, 23, 75, 121

Farm and Conservation Grant Scheme,
UK 135
Farm Diversification Grant Scheme
(FDGS) UK 131
Farm Woodland Grant Scheme UK 135
farms and farming 143–145, 160
see also agriculture; livestock
arable farming 5, 7, 13, 47, 60, 84,
98
diversification 103–119, 121–123,
127, 133, 137
family farms 1, 84
farm type 87, 88, 107–110
farmers 34
behaviour 136
as 'nature managers' 78
retirement and training 122, 135
urban fringe farmers 88
farming systems 26–27, 104, 105–112
income 5, 115
industrialization 4, 31, 91, 92, 122,
226
other gainful employment (OGA's)
127, 130
production 23, 25
quotas 105, 124–125
size distribution 107–108
subsidies 4, 6, 16, 122
surpluses 5, 23, 123, 136–137

technologies 3, 82, 88, 91, 92, 96,
122
fauna *see* animals
Fens 131, 177
fertilizers
see also pollution
artificial 195–196
input increase 133
manures, animal 7, 9, 98, 194, 231
nitrates 179–182, 185, 195–196
organic 202
phosphates 6, 141, 148, 200
pollution 6, 195–197, 199–200, 231
Ukraine 225, 226
use based on official data 165
Festuca ovina L. 241
Festuca rubra L. 241
Festuca sulcata (Hack.) Num.p.p. 237
fishing 47
flora 66
species 23, 55, 113, 114–115,
117–118, 177, 179
floristic gradient 114
Food and Agriculture Organization
(FAO) framework 207–208
forests and forestry 236–239
see also woodlands
coenoses 241
coniferous trees 182, 213–214
coppice biofuel production 137
deciduous 213
degradation 230
development 5–6, 59
land-use category 30
legislation 67
lethal irradiation of pine forest 240
new and removed 88
plant biomass 47
property rights 84
steppes 13–15
zones, Russia 211–212
frameworks
see also analysis and diagnosis
analytical 24, 27, 37, 81–83, 99–10,
104
biotic regulation potential 55
Environmental Basis Act (1987) 65
EULANU 15–17

Food and Agriculture Organization
(1976) 207–208
integrated transect method (ITM) 35
land-cover 24
land-use diversity 103–106
landscape 23–24, 43–46, 85–101
methodological 245–247
Regional and Municipal Master Plans
68–69
research 1, 43–46
Fraxinus exelsior L. 236

General Agreement on Tariffs and Trade
(GATT) 103, 123–125
General Agricultural Census (1988) 106
Geographic Information Systems (GIS)
modeling 209–210
Geographic Information Systems (GIS)
35, 208–209, 219
geography 36, 37, 43, 46, 218–219
geomorphology 29, 95, 96
German Democratic Republic (GDR)
141, 143
German occupation, Estonia 173–174
Germany 1, 6–7, 44, 125, 154
global climatic changes 25, 208
grant schemes 122, 129, 130–131, 135,
148
grasses 241
grassland
driving factor 109–110
effects of erosion on 197
management 6, 111–113
Molinia 194
mowing 10, 110
nitrogen input 16
permanent
agreements 99
importance of 116–117
land-use category 8–9, 86, 87
percentage of farm land under
106, 107
plant biomass 47
strips 14
grazing 99, 105, 110, 112
green zones 43
Gross Domestic Product (GDP) 206,
232

groundwater
contamination 7, 142, 146–148, 160,
185
level changes, dynamic aspects of 46
pollution 6, 179–182, 199–200
recharge 46, 49, 51
resources 99
table 88, 93
Guadiana river 72

habitat
animals 90, 148
biota 202
biotopes 61, 90
conservation 60, 160
development 16
game 93
great bustard 72, 75–78
loss 122, 177
plant 90, 148
protection 5, 66, 67, 75–76
recreational 161
restoration 123
value assessment 55–57
wildlife 17, 148
Habitat Improvement Scheme UK 129
Habitat Moorland Farm Woodland
Premium UK 130
Halle 9
Harz Mountains 154
heathlands 59, 84, 94, 99, 142
hedgerows 15, 105
avifauna breeding 10
cattle enclosures 90, 93
connective landscape element 17
networks 36
new and removed 26, 87–88
planting 84, 89
topographical maps 30–31
windbreaks 90, 93
Hercinic Massif 70
historical developments
abolition of serf-dom in Russia 172
Biskupan, Poland 191
Estonia 169, 173
Industrial Revolution 6
Iron Age 191
Middle Ages 6

Napoleon's campaign 192
Neolithic Age 190
Polish state formed 191
Soviet regime collapse 170
Ukraine 239
World War II 115, 121, 153, 165, 169, 174, 195
Homestead Board, Estonia 173
human activity 27, 70, 90, 105
influence 43, 53, 58
human resource management 26
human-ecological functions 48
Humberside 127
Hungary 221
hunting 47, 67
hydrology 12, 27, 47, 96, 193–194, 221–222, 226
hydromorphy 106–107
Hypericum perforatum L. 241

IGBP/IHDP Core Project on land use and Land-cover change (LUCC) 247
impact assessment 55, 66, 68, 122
Important Bird Area (IBA) 75–76
indices 166–167, 210, 211, 224, 225
industrial crops 7, 11, 150
Industrial Revolution 6
industry 141, 154, 185, 190
emissions 228–231
pollution 182, 222, 226
wastes 93
insect communities 200
Institute of Environmental Protection, Estonian Agricultural University 166
INTECOL Agrosystems Working Group 2
Integrated Administration and Control System (IACS) 124
integrated transect method (ITM) 35
intensification 25, 35, 60, 61, 70, 94, 112, 115, 117, 121
International Association of Landscape Ecology (IALE) 247
irrigation 71, 73, 88–89, 97–98
ditches 11, 24, 90, 93
Italy 125
Ivano-Frankivska region 226

Joint Research Centre of the European Commission, Ispra, Italy 17
Jutland 29, 91–98, 95

Kaluga 214
Kamenka 236
Kazakhastan 206
Kherson 226
Kiev 1–2, 231
Kirovohrad 226
Koeleria glauca (Sorend.) DC 237
Kostyantinyvka 229
Kriviy 229
Krivorizhzhya 231
Kuban river 21

Lahemaa National Park 183
lakes 14, 93, 94, 142, 167
Land Cadastre of the Estonian SSR 165
Land Cadastre of the Republic of Estonia 165
land redistribution 168–169, 173–174
land reforms 168–169
land unit, definition 207–208
land use 81–119, 221–244
agrosilvopastoral 14
categories 31, 84, 111
changes through drainage 193–194
driving forces 28, 81–102, 143–145, 167–172
management techniques 43–46
patterns 8, 9, 14, 33, 34, 37
planning 16, 28, 32
reforms 195
statistics 29, 33
sustainability 24
systems 23
tenure 144
land-fill sites 93
land-use 35, 103–119
land-use systems, defined 27
Land-Use Systems (LUS) 107
land-use types (LUT) 217
land-use/cover dynamics 215
landcover 237–239
landscape 103–119
analysis and diagnosis 1–64
changes through drainage 46, 177

development   4, 8–9, 10–14, 35,
   44–45, 46
diversity   9–11, 148
multiple use   155–160
networks, organisms, movement of   9
patchwork   14, 33, 36, 98, 107,
   172–174
pattern dynamics   82–102, 245–247
planning   48
   research   43–46, 54
   restoration   239–241
   structure   24, 87
   synthesis   44
   systems   24, 26–37
law and legislation   66–68, 183
Leipzig-Dresden School of Landscape
   Ecology   14
Leipzig-Halle   1, 141–163
Leipzig-Halle-Bitterfield agglomeration
   145–148
Less Favoured Areas (LFA)   122
linear landscape elements   14
Lisbon   66
Lisytchansk   231
livestock
   see also animals
   cattle   87, 106–118, 169, 171
   enterprises   4, 6–7
   farms   169
   feed   84, 107, 109
   grazing   74, 75
   headage payments   129
   numbers   86, 143–144, 160, 171
   pigs   12, 87, 169
   premiums   99
   production   91, 122
   Russia   206
   sheep   16, 87
   stocking densities   16
   wastes, treatment techniques   201
London   127, 129
Lugansk   222, 230–231

MAB project   1–21, 245–247
MacSharry proposal (1991)   124
Main Act on protected areas, Estonia
   65, 67
Map of Agricultural Regionalization   210
Map of Land Use   210

mapping
   agrolandscapes   217
   analysis   166–167
   digital ecological map of Europe   17
   geo-related maps   96
   geomorphological maps   29, 95
   Jutland   96, 97
   land-use maps   30–31, 165
   landscape, Chernobyl   243
   landscape and ecological maps   66,
      215
   thematical maps   55, 209–210
marginalization   25, 78, 81, 85, 95–99
   see also afforestation
Mariupol   229
marl pits   84, 90, 93
Master Plan, Castro Verde   65, 73
material flows   9, 11–12, 14, 246
Mazurian region   193
meadows
   alvar   185
   as a category   30
   duration of development   59
   ecosystems   198–199
   hay   113
   ploughing   54
   protection   94, 237
   sedge-hygrophytous   237
Mediterranean   6, 14, 32–33, 35–36,
   70, 72, 78
Meshera lowlands   213
meteorology   47, 48, 166
Military Historic Museum Reserve   217
mining   25, 73, 141–143, 149–151,
   222–232
Ministry of Agriculture   77, 117, 165
Ministry of Energy   229
Ministry of the Environment   77
Ministry of Planning and Land
   Management, Portugal   65, 66
modeling   10, 24, 25, 34, 218–219, 246
   see also analysis and diagnosis;
      frameworks
   functional models   105–106
   Geographic Information Systems
      (GIS)   209–210
   land-use dynamics   101
   plant succession   241, 242, 243
   RASTER   155

relating species   107
spatial aspects of agriculture   219
Molinia, grassland   194
montado   14, 73
Moritzburg   49
Moscow   214, 216
Motacilla flava   10
Mozhaisk   217
Mulde river   142
Munich   14–15
Munich School of Landscape Ecology   14
Municipal Council, Castro Verde   76
Municipal Master Plan (PDM), Castro Verde   68–69, 76, 78
Municipio de Castro Verde   73, 76
Mykolajiw, administrative districts   226

National Agricultural Reserve (R.A.N.)   67
National Ecological Reserve (R.E.N.)   67
National Land Board, Estonia   170
national parks   43, 66, 184, 196
natural land(scape) (NLU)   217
natural potentials/landscape functions   43, 44, 46–48, 233
natural resources   43–63
   exploitation   68, 69, 206–207
   management   27
   regulation   68
   water cycling   197
natural zones   211
nature conservation   5, 6, 24, 65–79, 131, 165
nature conservation and planning, tools   65–69
nature reserves   66, 184, 233
Netherlands   125, 232
Neviski biennis L   241
Nitrate Sensitive areas (NSA)   130, 135–136
Nitrate Vulnerable Zones (NSZ)   130
nitrates   179–182, 185, 195–196
nitrogen
   contamination   146
      decrease   160
   input to grassland   16
   leaching   32

pollution   6, 7
   surplus   15–16, 146–148
Nitrogen Sensitive Areas (NSAs)   136
Normalised Difference Vegetation Index (NDVI)   210, 211
Normandy   8, 34, 103–119
Numenius arquata   10
nutrients   37, 52, 165, 179–181

Obra river   193
Oentera biennis L   241
oilseeds   7, 11
Oka-Don Upland   214
Opachichi   236
Organic Aid Scheme, UK   129
organic farming   129, 135
organisms   9, 47, 55
Otis tarda   72, 75–78

Parnumaa   173
pastures 77, 84, 92, 94
Pays d'Auge   2, 8, 103, 106–107, 117
Perdix perdix   10
Perspectives for the CAP   123
pesticides   5, 6, 75, 133, 217
Phodococcum vitii-idaea (L.) Avror   237
phosphates   6, 141, 148, 200
physical environment
   bogs   59, 90, 92, 93, 167, 177
   drainage valleys   189
   fertile plains   25
   fluvioglacial sandy plain   235–237
   glaciation   167, 189, 235
   limestones   167, 177
   marshlands   193–194
   moraine   29, 179, 189, 236
   outwash plain   30–31, 84, 93
   palaeogenic clays   236
   peat bog   236
   river valley   29, 31
   salt marshes   94
   sandstones, Devonian   167
   tundra   211
   Ukrainian Crystalline Shield   235
Pinetum   237–239
Pinetum cladinosum   237
Pinetum phodococco-dicranosum   237
Pinus silvestris L.   236

plant   107
  see also habitat; species
  associations   70
  communities   56
  cover structure   200
  growth season   11
  populations   142
  succession   241, 242, 243
plantations   31, 34, 71, 74, 76
Poland   1, 10, 11, 189–204
Polesie   235, 236–237, 239
policies
  agriculture   4, 6, 29, 84, 122, 141–145
  agro-environmental   84, 94, 99
  development   5, 13, 27, 33
  driving force   82–83
  environmental   65–79, 216–218
  incentive schemes   135
  national, Denmark   98
  planning   99
  regional   66
  Russia   206–207
Polissya region   233
political issues   168–171, 174,
  184–185, 196
pollutants   232
  nitrogen   6, 7, 111, 112, 114, 141
    concentrations   179
    contamination   146, 160
    decrease   160
  nitrogen oxides   232
  nutrients, run-off   37, 52, 165,
    179–185
  pesticides   5, 6, 75, 133, 217
  phosphorus   226
  slurry disposal   7
  spatial spread of   199
  sulfur oxides   225
pollution   141–151, 179–182, 235–243
  see also contamination; fertilizers;
    pollutants
  atmospheric   226–230
  control   201, 202
  emissions   226–227
  environment   12
  fertilizers   146, 160, 225, 226
  groundwater   6, 151, 199–200
  industry   222
  reduction of   15

river   24
soils   23
Ukraine   229–230
water   16, 23
ponds   12, 14, 87, 88, 90, 93, 200
population   68, 190–193
  density   185
  deportation   169, 173–174
  distribution   72–73
  growth   6, 193
  impact of   190–193
  rural areas   76–78, 141
  urban   137
Porijõgi river   165–187
Portugal   1, 3, 14, 65–79
post-productivism transition   125–126,
  131–136
Potentilla argentea L.
power
  generation   225, 232
  stations
    biomass fuelled   15
    dominating land use   13
    emissions   229
    Hydroelectricity power stations
      (HPS)   231
    Leipzig-Halle   150
    nuclear power plant (NPP)   235,
      240–241
    reservoirs   231
    water use   230–231
Poznan   190, 193
present day landscapes (PDL)paradigm
  207–210
Prichornomoyre region   233
productivism transition   121–139
Prosna river valley   194
Provence   32
Prydniprov'ya   229
Prypyat   236–237

Querceto-Pinetum   238
Quercus ilex   74
Quercus robur L.   236, 242
Quercus suber   74
quotas, milk   106, 124–125

reclamation   94, 149–150, 177, 180,
  185, 227

recreational areas
  accessory land use   13
  biotopes   90
  diversification into   131, 137, 183
  EU reports on   78
  lakes as   151
  landscape preservation for   160, 233
  potential   46
  urban fringe   85
Regional Coordination Commission
  (CCR)   66, 76
Regional Plans (PROT)   68–69
regional policy, Russia   206–207
regulatory instruments   23–25
remote sensing   17, 103, 206–212, 218
Reola hydrological measuring point
  166
reprivatisation of land   165, 170
Research Centre for Agricultural and
  Forest Environment   197
reservoirs   12, 199, 231
Rig   229
river pollution   24
Romania   7
run-off regulation   46
rural areas   23–42, 72–73, 76–78, 141
Russia   1, 7, 205–220

Saale river   142
Saaremaa   173
Sado river   70, 72
Salix acutifolia Willd.   237
satellite images   55, 209, 212–213
Saxonian School of Landscape Ecology
  44
Saxony   46, 62, 141–163
Saxony-Anhalt   7, 155
School of Landscape Ecology, Munich
  14
set-aside
  arable enterprises   7
  Copenhagen   86
  decrease in biotope value   60
  permanent   99
  Saxony   143
  under CAP   25–26, 99, 160
  United Kingdom   127–129
Setaria glauca (L.) Beauv.   241

settlements   14, 151–153, 239, 246
shelterbelts   11, 94, 197–201
  General Dezydery Chaplowski   193
Siberia   210, 211
Silversky Donets river   231
Slovakia   1
soils
  characteristics   72
  compaction   47, 150
  contamination   159, 237
  degradation   14, 194, 201
  erosion   13–16, 23, 161
    cereal cultivation   70–71, 98
    control   192–193, 202
    water and wind   11, 31, 51, 191,
      197, 226
    resistance to   46
  loss of fertility   191
  material flows   11–12
  pollution   23
  water movement   12
South Estonian Laboratory of
  Environmental Protection   166
Soviet Union   4, 12, 169, 173–174, 216
spatial analysis
  biotope evaluation   60
  frameworks   69
  land-use strategies   35
  patterns   85, 95, 103, 118
  resolution   32, 218
  scales   32, 36–38, 48, 81, 88, 121
  structures   48, 104
  variations   93
species
  see also animals; birds; flora
  colonization   17, 105, 115
  dispersal   24
  diversity   10, 16, 57
  endangered   11, 61
  grasses   241
  loss of   23, 177, 179, 185
  mobility   10
  plant   10, 54, 55, 60, 193
  protection   66
  rare and endangered   11, 61
  sampling of   107
  trees   74, 77, 236–238
  vegetation   79, 177

Species and Habitat protection 67
spiders 107
standard gross margin (SGM) 125
state farms 4, 170, 206
state intervention 1, 105, 121–138
State Land Reserve, Estonia 168
State Nature Reserves, Estonia 184
State Treasury Agricultural Property
    Authority (STAPA) 196
statistics 33, 37, 103, 115–116, 143,
    165–166
subsidies 4, 6, 16, 94, 122
Sumy 226
surveys 44, 55, 107–109
sustainable agriculture 26, 65, 121,
    201, 232–233

Tagus river 70
Tartu-Ulenurme Meteorology Station
    166
Tcherkasy region 226
Tchernihiw region 226
technology 29, 30, 52, 93, 96, 99–101
Tejo valley 66
test areas 54, 55, 58, 92
    England 131–136
    Estonia 174–177
    Moritzburg 49
    Moscow 216
    Russia 216
    Saxony 62
    Tver 216
The Future of Rural Society 78
Thuringen, Germany 16
timber 13, 84, 131, 214
time 10, 34, 35, 44, 50–53
tourism 68, 78
traffic 69, 141, 153, 154
transition problems in Russia 216–218
transport 25, 221–222, 226–228
Trebon 1
trees 15, 74, 77, 236–238
Tula 213
Tune area 88
Turew 193
Tver 213, 216
typology 36, 126

Ukraine 1, 2, 12–14, 13, 206, 221–234

UNESCO 1
United Kingdom 1, 121–139
urban
    fringe 85–88, 90, 169
    growth 24, 68, 81, 85, 99–100
    land 17, 160, 222
    pattern 69
    spread 215
    transport 222

Valday Upland 213
Vale of Evesham 129
vegetation
    characteristics 10, 167
    cover 14, 35, 58, 90, 207, 216
    distribution 70
    diversity 56, 59
    indices 210, 211, 212
    influence on water cycling 11,
        197–199
    mapping 55
    regeneration 46, 47, 56, 61, 241
    riparian 74
    sampling 107
    spatial criteria 56, 58, 60
Vinnytsya region 226
Virumaa 173
Vladimir 212, 214
Volga river 21
Volgograd 214
Voloshyn, V. 2
Volyn region 226
Vorumaa 173

Wales 129
wars 6, 115, 121, 153, 165, 169, 174,
    195
Warta river 189–190, 191, 193
wastes 93, 182, 222
water
    accumulation 47
    balance 46, 154–155, 156, 158
    contamination 6, 159, 160
    cycling 11, 12, 197–199
    improvement 13
    legislation 67, 68
    management 198–199
    pollution 23, 122, 135, 151
    protection 15, 160

resources   74, 97, 197–199, 201, 230–231
  surface waters   46
watersheds   12, 37, 38, 190
Weiße Elster river   142
Wielkopolska   11, 189–204
Witham river   131
Wolds   131–136
woodlands   236–239
  see also forests and forestry
  agroforestry   77
  duration of development   59
  Leipzig-Halle   155, 157

montado   14, 73
  reduction   70
  revitalization   6
World War II   115, 121, 153, 165, 169, 174, 195

Yaroslavl   214
Yearbook of the Estonian Land Cadastre   165
Yearly Cadastre, Estonia   170

Zaporizhzhya   229, 230
zonation schemes   89–90, 211